Stress Agronomy

The Author

Professor (Dr.) N.K. Prasad is a former Professor (Agronomy), BAU, Ranchi and Professor (Agriculture) in Ethiopian Universities is the recipient of the Common Wealth countries fellowship sponsored by the Reading University, UK for Post Doctoral Research and Training on "Micronutrients nutrition to a series of legumes" at CSIRO, QLD., Australia for his outstanding research achievement on Lucerne production in acid soils of India. He is also the recipient of 'The "Best Paper Award" of the Indian Society of Agronomy, New Delhi. He has published about 100 research papers in different National and International Journals and guided several M.Sc. students in India and abroad encluding Ph. D. Scholars of Agronomy and Soil Sciences in India. Stress Agronomy is the second Book authored by him after first one on "Soil Fertility and Plant Nutrition".

Stress Agronomy

— Author —

N.K. Prasad

Post Doctorate (Australia)
Ex Professor, Agronomy (BAU), Ranchi &
Professor, World Bank Projects, Ethiopia

2014
Daya Publishing House®
A Division of
Astral International Pvt. Ltd.
New Delhi – 110 002

Published by   : **Daya Publishing House®**
                            A Division of
                            Astral International Pvt. Ltd.
                            – ISO 9001:2008 Certified Company –
                            4760-61/23, Ansari Road, Darya Ganj
                            New Delhi-110 002
                            Ph. 011-43549197, 23278134
                            E-mail: info@astralint.com
                            Website: www.astralint.com

Laser Typesetting   : **Classic Computer Services**, Delhi - 110 035

Printed at   : **Replika Press Pvt. Ltd.**

PRINTED IN INDIA

Acknowledgement

Crop management stratedgy under changing climate and problematic soils conditions is the call of the day. Balloning of population and shrinking of per capita culturable land has hardly any option left for horizontal expansion for crop husbandry. Hence technology is in utter demand to manage the crops under soil and climatic stresses to feed more than 1.2 billions people of the country and over 7.0 billion world population.

Therefore attempt has been made to put the matter before the screen for possible technologies to coup with the burning issues. The author acknowledges the encouragement and suggestions given by Dr. R. Thakoor, Prof. Agronomy, Dr Balraj Singh, Prof. Animal Sciences and other friends. Mr. Nrip Kishor "Neeraj" deserves special appreciation for his non-stop cooperation in preparing the figures, graphs, tables and proper arrangement of the manuscript.

The information gathered and incorporated in this book from different books and publications of authors as referred over here is warmly acknowledged.

N.K. Prasad

Preface

Swelling of human population and shrinking of cultivated land area has led to availability of only a few meters of agricultural land to per capita (about 1200 m^2/capita) in India. This has compelled us to bring the problematic soils under the plough either through soil amendment or by selection of resistant genotypes through breeding to meet the food demand. In addition to this, the exploitation in population and its demands for day to day industrial products has led to pollution of soil and climate to such an extent that survival of several crop plant species in coming future might be suicidal for all the creatures on this living planet.

Therefore, the world scientists are trying to cope with the changing climatic conditions since last 3 decades. Human activities have virtually over exploited the natural resources to meet their growing demands. Emission of Green House Gasses (GHGs) has direct effects on increasing in temperature resulting into melting and shrinkage of glaciers and thus rising in sea-levels with incidences of unprecedented flood and drought. It has poised a challenge to agricultural scientists throughout the world to evolve the new genotypes of crop varieties and appropriate technology to grow and feed the swelling population. Occurrences of new diseases are further a challenge to the health sector due to pollution of climate, water and soils in general and heavy metal toxicity in particular.

Heavy industrialization in general and power industries in particular as well as rising in demands for petroleum products has toxified the atmosphere so badly to the extent that even it is difficult to enjoy a healthy environment. The warming in earth system will virtually change the ecology of the whole world. There may be transformation of tropical countries in semi-arid while temperate world may change to sub-temperate or sub-tropical. Therefore, the farming systems may be altered universally. It is high time for the International organization to resolve the pollution issue without accusing each other. If rice growing developing Asian countries are responsible for methane emission, the developed countries are more responsible for emissions of several poisonous gases and other agents of pollution which have longer life span in the atmosphere than methane.

This book is aimed to address the past, present issues together with expected changes in soils and climate in future, The thrust is aimed to meet the challenges through evolution and adoption of new technologies in agriculture sector to facilitate the coming generations to avail a pollution free life. The information gathered here may be an eye opening for agriculturists, environmentalists, climatologists, planners and policy makers to take knowledge of present day burning issues of atmospheric, terrestrial, oceanic changes and global warming due to emission of poisonous gases on one hand and heavy metal toxicity in soil and air on the other. Hence, the scientist community and policy makers have to bring out balance between human population and their requirement without extreme exploitation of the natural resources for a sustainable eco-friendly system. If, not, global warming is a worth warning which may prove to be a global harming due to possible change in bio-planet to a non-living one.

The author is grateful to different authors' publications from where the vital information are collected and incorporated in this book for the benefit of the students, scientists of various disciplines for future research and for the planners and policy makers to give upmost thrust on this line.

N.K. Prasad

Contents

Acknowledgement   *v*

Preface   *vii*

Abbreviations   *ix*

1. Introduction   1

Part I–Soil Stress

2. Soil Acidity   9

3. Soil Alkalinity   18

4. Soil Salinity   27

5. Flooded Soils   35

6. Drought Stress   44

7. Agronomy of Drought Stress   57

8. Nutrient Stress   67

9. Water Requirement   77

Part II–Climate Stress

10. Population and Pollution   89

11. Soil and Water Pollution   97

12. Light Energy Harvest   111

13. Photosynthetic Stress   121

14. Respiratory Stress   143

15. Abiotic Stress and Crops   148

16. Biotic Stress and Crops   166

17. Mitigation of Polluting Agents   173

    References   181

    Index   183

Abbreviations

ABA	–	Abscisic acid
Am	–	Arbuscular micorrhizal
C_3	–	3 Carbon compound
C_4	–	4 Carbon compound
CAM	–	Crassulacean Acid Metabolism
CFCs	–	Chloroflorocabons
DNA	–	Deoxyribo Nucleic Acid
DW	–	Dry weight
ENSO	–	El Nino Southern Osciliation
ET	–	Evapo-transpiration
GHGs	–	Green House Gases
IAA-	–	Indole-3-acetic acid
LAI	–	Leaf Area Index
MA	–	Mugenic acid
Mpa	–	Matric Potential (Pascal pressure a unit of 1 Newton/ m^2)
Nm	–	Nanometre ($1 \times 10^{-9} m$)
OPPP	–	Oxidative pentose phosphate pathways

PAR	–	Photo active-radiation
PEPC	–	Phosphosphoenalpyruvate Carboxylase
PCR	–	Photosynthetic Carbon Reduction
PGA	–	Pyruvate glyceric acid
PQ	–	Pastoquinone
PS I	–	Photosystem I
PS II	–	Photosystem II
Q10	–	Quantum jump/doubling in yield to per 10°C rise in temperature
QA	–	Quinine acetones
RGR	–	Relative Growth Rate
Rubisco	–	Rivulose 1, 5-bisphosphate carboxylase oxigenase
RQ	–	Respiratory Quotient
RuBP	–	Ribulose- biphosphate
SAR	–	Sodium Absorption ratio
SIC	–	Soil inorganic carbon
SOC	–	Soil organic carbon
SOD	–	Soil Oxygen Demand
UB-A	–	Ultraviolet-A radiation
UB-B	–	Ultraviolet-B radiation
UB-C	–	Ultraviolet-C radiation
VPD	–	Vapour Pressure Deficiency
WUE	–	Water-Use-Efficiency

Chapter 1
Introduction

Stress' or 'pressure' has been introduced into the theory of elasticity as an amount of force for a given unit area. When sufficient force is applied to material, the material bends and the change in length is termed as 'stress'. In short term 'stress' is the action whereas 'strain' is the reaction.

In terms of crop science, it is the plant stress caused by abnormal conditions in soil and climatic conditions for normal plant growth and development process which is the reflection of the amount of environmental 'pressure' for the change placed on plant growth due to changes in physiology while proportional change in yield performance as a consequence of 'stress' is termed as 'stress. Lachtenthaler (1996) defined plant stress as any unfavourable condition or substance that affects or blocks the plant metabolism, growth or development, by Strasser as 'a condition caused by factors that tend to alter the equilibrium' while according to Larcher, it is the changes in physiology that occur when species are exposed to extraordinary unfavourable conditions that need not represent a treat to life but will induce an alarming response.

In plant science, stress has been classified into two groups as 'biotic' which is caused by living organism (insects, diseases and herbivores) and 'abiotic' caused by natural incidences (flood, drought, problematic soils conditions, pollutant, heavy metals and others) are illustrated as under (Figure 1.1).

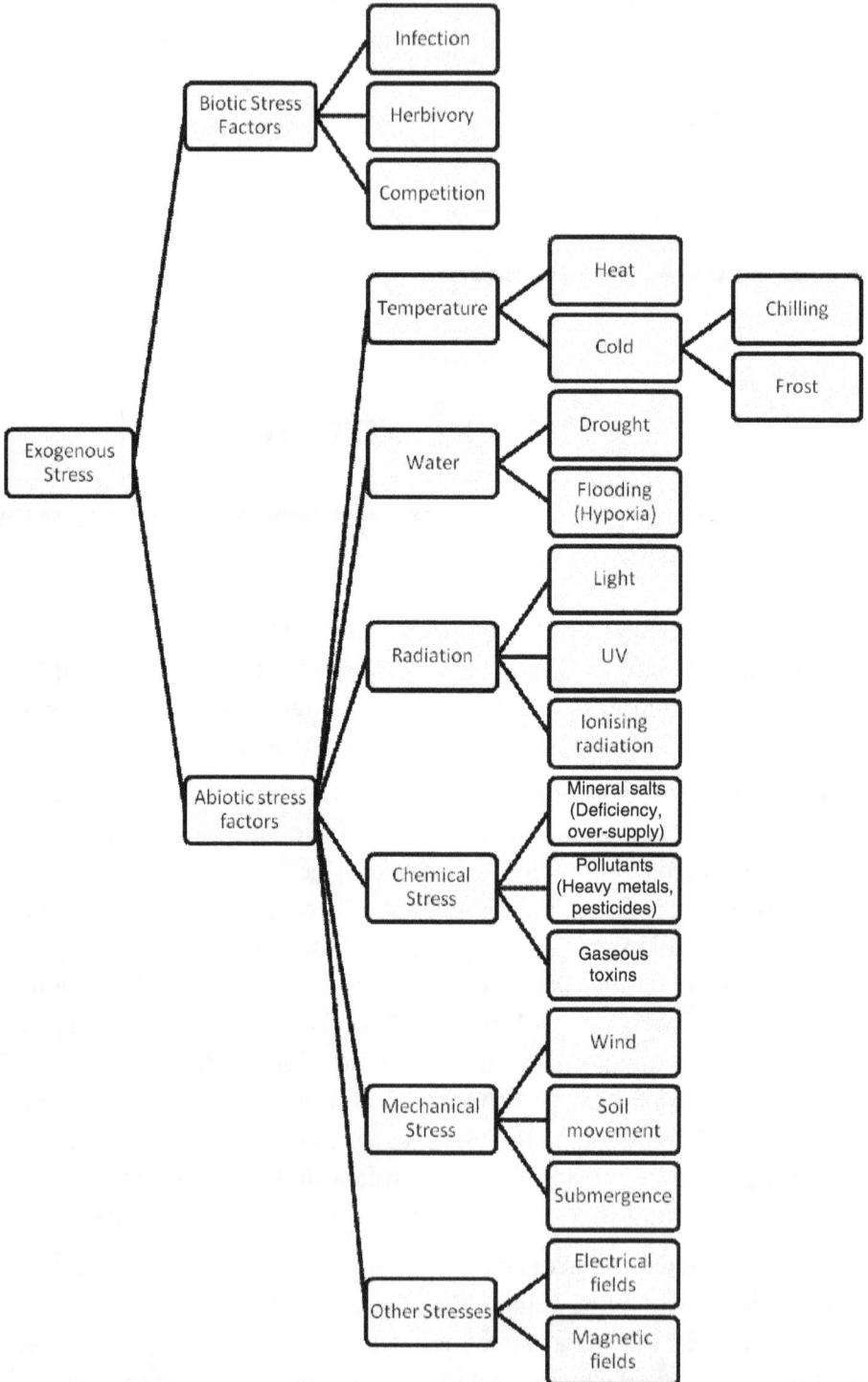

Figure 1.1: Biotic and Abiotic Environmental Factors Causing Stress for plants.

Almost similar to this classification, some other scientists have further grouped the abiotic stresses into physical, chemical and anthropogenic stresses (Table 1.1).

Table 1.1: Classification of Stress.

| | Biotic | Abiotic | |
	Physical	Chemical	Anthropogenic
Pathogens	Flood	Soil pH	Air pollution
Herb ivory	Drought	Salinity	Pesticides
Allelopathy	Radiation	Nutrients	Heavy metals
Competition	Temperature	Soil organisms	Fires
Mycorrhizae	Wind	Atmospheric gasses	Alieniwasin

Any stress conditions to plant does not result only obstruction in plant physiological process but has direct bearing on several other biological functions to put the plant under stress conditions and subsequent reduction in yield, accordingly. Beside other factors, soil and climatic conditions determine the types of plant species to be grown in a particular region. Therefore, any abnormal soil conditions and changes in climatic parameters bring out changes in the crop and cropping system of the area in question. Hence crop management in problematic soils under stress climatic conditions is termed as stress agronomy. As Agronomical point of view these stresses may be listed as under:

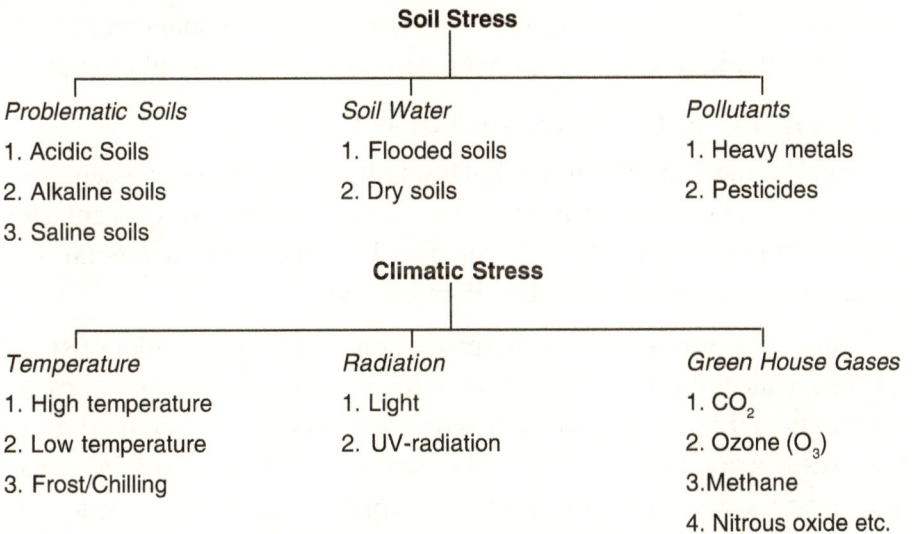

Soil Stress

Problematic Soils	Soil Water	Pollutants
1. Acidic Soils	1. Flooded soils	1. Heavy metals
2. Alkaline soils	2. Dry soils	2. Pesticides
3. Saline soils		

Climatic Stress

Temperature	Radiation	Green House Gases
1. High temperature	1. Light	1. CO_2
2. Low temperature	2. UV-radiation	2. Ozone (O_3)
3. Frost/Chilling		3. Methane
		4. Nitrous oxide etc.

Management of crops and cropping systems under problematic soils and polluted climatic conditions is a great challenge for the Agronomist to harvest the optimum yield to feed the world swelling population in near future. The rise in temperature in general and especially in tropics and subtropics due to drastic depletion in soil organic carbon and subsequently increases in atmospheric CO_2 concentration and temperature may alter the plant ecology of different regions and as such the crops and crop production system. There is every possibility that area under problematic soils may increase due to the erratic behaviour of the climate due to pollution, some area may receive more than excess rain to create flooding on one hand, while some other area may experience drought like situation lead to increases in arid land on the other. Increases in population will further toxify the soils, water and air to create human health hazards. Increases in temperature due to burning of soil organic carbon may increase the CO_2 concentration of the atmosphere which will further increase the temperature. Existence of mutual relation between CO_2 and temperature will reduce the thickness of the ozone layer to allow UV-B radiation to pass on the earth surfaces to help increases in atmospheric temperature. Increases in living human demands are further contributing increases in the concentration of CFCs to create further injury to vegetations.

Since plant is the only industry that transforms solar energy into food material for the survival of all types of lives on this planet hence attempt has to be made for the continuity in sustainable crop management system under problematic soils and climatic conditions for survival of mankind.

1.1 Adaption of Plants to Soil Stress Conditions

Soil chemical conditions (soil pH, salinity and nutrients concentration) and climatic conditions (temperature, rainfall pattern and concentrations of green house gases) determine the distribution of natural vegetations in different ecological zones on this living planet.

Generally crop species are adapted to high fertility conditions (Ruderal species) while wild species are adapted to low soil fertility are slow growers usually have high root shoot ratios and low maximum potential growth rate. These grow slow even with the supply of nutrients but are efficient in storing the energy to use when the supply is restricted. Chapin (1988)

Nutrient Availability	Type I (Slow growers)	Type II (Ruderal species)
Low (nutrient poor sites)	Low nutrient uptake rates	Low growth rates
	Low growth rates of roots and shoots	Low nutrient storage
	High leaf longevity	High root/shoot ratio
	High nutrient concentration in the tissues.	
High (nutrient rich sites)	Small growth response of roots and shoots	High nutrient uptake rate High growth rate
	High nutrient storage	High NUE
	Low nutrient use efficiency	Decrease in root shoot ratio.

Figure 1.2: Strategies of Plants in Adaptation to Soils with Low or High Nutrient Availability.

classified the natural vegetation into two types (Figure 1.2) on the basis of growth habit as slow growers (Type I) and fast grower at high fertility conditions as ruderal species (Type II).

Large seed size containing high nutrients for seedling development in poor soils is also one of the characteristic of plants under Type I

Plants like Artic sedge (*Eriphorum vaginatum*) are efficient in utilization of organic soil nitrogen either by hydrolysis of proteins or preferential uptake of amino acids under limited nitrogen conditions in temperate climate.

Under P- limited conditions, role of VAM in P nutrition is the same for wild and cultivated plants. Very high P- efficiency in wild oat (*Avena sativa* L.) is due to more developed root system.

1.2 Soil Stress/Problematic Conditions

1. Acid Mineral Soils
2. Alkaline Soils
3. Saline Soils
4. Flooded and waterlogged Soils
5. Drought conditions

PART I
Soil Stress

Chapter 2
Soil Acidity

Almost 70 per cent of world vegetations are found in acid soils which indicate the adaptability of majority of species in acid environment despite the following major constraints.

1. H+ toxicity due to high H+ ion concentration.
2. Al- toxicity   H+
3. Mn- toxicity   H+
4. Decrease in macronutrient cation concentration : Mg, Ca and K deficiency.
5. Decrease in solubility of P, Mo and B.
6. Restriction in root growth resulting in low water and nutrients uptake efficiency and leaching of nutrients.

Continuous growing of N- fixing legumes, Mn^{2+} toxicity may develop in high Mn^{+2} containing acid soils. High concentration Al^{+3} and to some extent of Mn^{+2} concentration are the major causes of stress in growing crops in acid soils (pH<5).

2.1 Solubility of Aluminum and Manganese

Aluminum in acid soils (pH 5.5) acts as a strong absorber of Phosphorous and Mo since clay mineral is occupied by Al as it replaces divalent cations (Ca^{2+} and Mg^{2+}) and hence root growth of several plant species is restricted. The effect is reduced by the formation of complex ions with organic and inorganic compounds due to the fact that entry of these big size complex ions are restricted in xylem vessels.

At low pH, the amount of exchangeable manganese is also increased which is a function of redox potential ($MnO^2 + 4H^+ + 2e \rightleftharpoons DMn^{2+} + 2H_2O$). Soil pH is also reduced by symbiotic legumes in soils with high level of readily reducible manganese. As such a pasture dominated by *Trifolium subterraneum*, soil pH is reduced from 6.1 to 4.8 and exchangeable manganese increased by 10 times (from 4.6 to 46.1 mg/kg soil) There is a less risk of manganese toxicity than of aluminum toxicity in highly weathered acid soils of the tropics.

Nodulation in acid mineral soils is adversely affected by a combination of high Al or Mn or both and low Ca concentration which is also adverse for P- nutrition and nodulation.

Aluminum Toxicity and Phosphorus Uptake Stresses

Stresses caused by toxicity of aluminum resulted in restriction in phosphorus nutrition of plants in acid mineral soils.

Aluminum Toxicity and Plant Stress

Aluminum is the third most abundant element in the Earth's crust (Average 82 g/kg soil) within the range of 10 to 300g/kg soil. 30 per cent area of top and sub-soils and 75 per cent of the total ice-free land area of the world are acidic of which Acrisol contributes the maximum area. Al-solubility increases as the soil pH decreases to 5.0. At pH 4.0 both Al and H+ are toxic. Al-toxicity can be predicted by measuring soil exchangeable-Al with1-M KCl. Al-toxicity in common crops occurs when exchangeable-Al > 2 cm $mole_c$/kg and even >1 $cmol_c$/kg is toxic for sensitive crops.

Mechanism of Al-Toxicity

Al-toxicity has direct effects on root growth, particularly on root apex and elongation. Inhibition in the growth of meristems by Al results in restriction in cell division. Since Al^{3+} has a high binding affinity to plasma membrane and shows a 500-fold higher affinity for the phosphatidylcholine surface than Ca^{2+} and therefore cells become more leaky and rigid to Al binding to plasma membrane which changes both the membrane potential as well as surface potential (Zeta potential) to restrict nutrient uptake.

Since, Al increases the molecular weight of hemicelllose and amount of wall-bound ferulic and di-ferulic acids and hence Al-modifies the metabolism of cell wall and makes it rigid to inhibit growth. Al-toxicity further inhibits Ca-uptake and translocation even it replaces Ca^{2+} from root tip at low pH.

Mechanism of Al-toxicity Stress Tolerance

Plants have two different mechanisms to sustain under Al-toxicity. These are grouped into excluders: one which exudates organic acids to form complex-ion to restrict the entry of such ions into the xylem vessels due to increase in size of the complex ion as compared to simple ions while encluders allow the plants to tolerate Al-concentration in their root and shoot symplasm. Root of different plant species secret different organic acids to complex the aluminum.

Table 2.1: Organic Acids Exuded by Plant Species to Complex the Al-ions.

Plant Species	Organic Acids Exuded
Wheat	malate
Soybean, Snap bean and Cassia	citrate
Maize and Rye	malate and citrate both
Taro and Buck wheat	oxalate

Presence of Al-activated anion channel on plasma membrane has a greater role in binding or complexing the Al to restrict pass on the xylem vessels.

Genetic Improvement for Al-toxicity Stress Tolerance

Some Al- inducible genes have been detected but their functions are yet to be understood. Over- expression of citrate synthase isolated from *Pseudomonas aerugenosa* in tobacco and papaya plants enhanced secretion of citrate and complex Al- ions stress. Transgenic barely and suspension cell of tobacco introduced with wheat ALMT-1 gene exuded malate to bind Al and allowed normal growth. It confirmed that ALMT 1 gene, which encodes malate transporter and is triggered by Al, is capable to make plant tolerance to Al-toxicity.

2 2 Phosphorus Deficiencies and Plant Stress

Phosphorus in the Earth's crust ranges from 0.035 to 5.3 g/kg of soils with an average of 1 g/kg while plant requires 2 g/kg of P for optimum growth. Alkaline soils having more than 1 per cent organic matter, P-availability is determined by the desorption and adsorption of P. P – deficiency in alkaline soils is mainly due to very low levels of total P followed by moisture stress. In plants, P exists in several forms as sugar-P, nucleic acid, nucleotides, phytic acid (inositol-P), phospholipids, co-enzymes and plays vital role in energy storage and structural integrity. Only less than 20 per cent of applied P is absorbed by the plants and rest is fixed as Al-P and Fe-P in low pH soils and as Ca-P and Mg-P compounds in high pH soils. Severe deficiency of P occurs in acidic soils dominated by 1:1 clay minerals, particularly those with loamy surfaces in presence of oxides and hydroxides of iron and aluminum.

In soils, 15-80 per cent of total P occurs in organic forms. In acid soils, organic-P exists as adsorbed form in iron and aluminum minerals and adsorbed inorganic-P and are not mineralized by phosphatase enzyme. Water soluble organic-P is directly absorbed by plants but inorganic-P is taken by the plants after mineralization in presence of phosphatase enzymes. Bio-availability of soil-P depends on its chemical forms and is controlled by dissolution/precipitation of soil-P minerals, sorption/ desorption of sorbed P, and mineralization/immobilization of organic-P, respectively.

P-deficiency is the major constraint to limit the plant growth in acid soils of tropics and subtropics. Among several plants, Cassava (*Manihot esculenta* Crantz) an annual root crop has the highest tolerance capacity in acid soils as compared to sweet potatoes, yams and potato. Among legumes cowpea and peanut as well as a large number of pasture grasses (*Brachiaria* sp., *Pennisetum* sp., *Andropogon* sp. and legumes (*Stylosanthes* sp., *Cassia* sp. and *Macroptillium atropurpureum*) are very tolerant to acid soils. This is the fact that majority of world pastures are managed in acid soils. Differences in low pH tolerance also vary from one variety to another variety of the same crop plants in cases of both grain as well as forages.

Mechanism of P-acquisition in Plants

Different types of strategies are developed by the plants for absorption of phosphorus from P-deficient or unavailable soil-P.

1. Alternation of root architecture

2. Secretion of low- molecular-weight organic acids

3. Secretion of phosphatase enzyme and

4. Enhanced role of P-transporters

Alternation of Root Architecture

Some plants like *A. thaliana* develops vigorous root under P-deficient conditions to acquire more P while some other plants form **proteoid roots** (cluster of short lateral roots) in response to P-deficiency to get more contact with larger soil volume. Root architecture is also developed with application of some hormones like auxins for lateral root development, root hair elongation and density. In addition to auxins, ethylene and cytokinins are also found effective for better root architect formation.

Secretion of Organic Acids

Secretion of citrate and malate acids by plant roots like white lupin under P-deficient conditions dissolves Al-P, Fe-P and Ca-P for self utilization. P-induced secretion of organic acids has also been found in a number of pasture legumes including Lucerne (citric, malic and succinic acids), rape (citric and malic acids), rice (citric acid) and several others (malonic, oxalic, citric, malic and piscidic acids) by pigeonpea and pea roots to help release of low-available-P in soils. Besides increases in P-availability, secretion of citrate also helps in anion-transport.

Secretion of Phosphatase Enzyme

Roots of the plants like, white lupin also secrets phosphatases (S-APases) to hydrolyze organic-P compounds in the rhyzosphere and supply inorganic-P to plants under P-deficient soil conditions.

Enhanced Expression of P Transporters

Plants take up maximum P when $H_2PO_4^-$: HPO_4^- ions (inorganic orthophosphate) are in equilibrium. It is generally supplied by diffusion on the plant roots Since P concentration in plant is much higher than external solution hence P is actively transported against the concentration gradient. Two types of transporters: first is a high affinity transporter with a Michaelis constant (Km= 3- 10 µM) and second is a low- affinity transporter with a Km= 50-300µM. The activity of these transporters is increased under P- deficient conditions and subsequently an increase in uptake of inorganic orthophosphate occurs.

Genetic Improvement

Some improvements in P- acquisition characteristic in plants have been made through gene manipulation. As such, in *A. thaliana* high- affinity Pi transporter gene (Pht 1) transfer in tobacco, increased in P uptake to 3 times and double in growth under low P conditions. In Lucerne, over expression of MDH gene resulted in an increase in P acquisition. In rice, a novel transcription factor (OsPTF1) with a basic helix- loop- helix domain for tolerance to orthophosphate deficiency has been also identified. The transgenic rice, gave more number tillers, higher root- shoot biomass and P concentration in plants almost each 30 per cent higher than wild type plants in P- deficient conditions.

Some works on salinity resistant plants to adopt osmotically by increasing intracellular concentration of low- charged low- molecular-weight solutes (osmolytes) such as proline and others have been clarified to sustain under mineral stress conditions in saline soils.

2.3 Tolerance vs Avoidance

☆ Aluminum Tolerance

☆ Manganese Tolerance

☆ Nutrient Efficiency

Tolerating or avoiding the stress with varying degrees is the most important factor for a plant to sustain under acid soils conditions which can be summarized as

Adaptation

Tolerance of Stress		Avoidance of Stress
High tissue tolerance to sites, toxic mineral elements (Al, Mn) ' Includers'	☆	Exclusion of Al and Mn from sensitive
Low internal nutrient demand	☆	Exclusion from uptake by roots induced charges in the atmosphere
Effective re-translocation and compartmentation of nutrients	☆ ☆	High efficiency in nutrient acquisition Favourable rhizosphere microorganisms, associations myorrhizas

Plants of early fossil history of *Proteaceae* family are the best include/ accumulators of Al in their system. As such tea plants may contain up to 30 mg aluminum/g of dry old leaves. Presence of aluminum (=100 mM.

Al) induces tea plant growth as well as to calcifuge species *viz. Deschampsis fleruosa* and *Arnica montana* L. Many plant species tolerate the Al mainly by exclusion from sensitive sites in roots or from uptake in general by root- induced charges in the rhizosphere. Thus exclusion mechanisms may be due to

1. Exclusion from sensitive sites
2. Rhizosphere pH and
3. Complex ion formation with root exudates

1. Exclusion from Sensitive Sites

Cytoplasm, plasma membrane, apoplasm and the peripheral root cap cells of the plant are the sensitive sites for aluminum toxicity. Though, some scanty information are not enough for such process however, differences in the plasma membrane surface potential and binding of aluminum at the plasma membrane may be the possible reasons for variations in tolerance for excessive aluminum among plants Some works also suggest differences in CEC of roots may be also responsible for Al tolerance but it is still not confirmed.

2. Rhizosphere pH

At low soil pH (4-5) an increase in pH though decreases the concentration of Al $^{3+}$ but it may increase the phytotoxicity of Al. However, an increase in rhizosphere pH may decrease both exchangeable aluminum as well as the release of insoluble -Al to soluble Al. Increasing the rhizosphere pH, H$^+$ toxicity can be reduced and binding of Ca^{2+} and Mg^{2+} in the root apoplasm can be increased.

3. Root Exudates

Mucilage of cell and organic acids play a vital role in binding and complexing aluminum but higher concentration of aluminum further decreases the mucilage production. The mucilage secretion from root cap is ceased at 20 μm aluminum concentration in sensitive species but at 400 μm in the tolerant one. As such high mucilage secretion from the roots of *Aristida juniformis* may be responsible for high aluminum tolerance in natural grassland of acid soils.

Complex ion formation by organic acids, usually oxalic acid protects from adverse effects of free aluminum on root growth on one hand and facilitates phosphorus absorption on the other. Release of citric acid by bean roots may be responsible for Al – tolerance in acid soils. In case of *Eucalyptus species* formation of complexes of Al with **polyphenols** leached from the leaves may facilitate acquisition of P from extremely P- deficient acid soils.

2.4 Mycorrhizas

The role of mycorrhizas in P nutrition in inherent P- deficient acid soil is warranted when development of root system is restricted by Al- toxicity as well as exudation of organic acids by the root is obstructed. In tropical pasture legume (*Stylosanthes guianensis*) role of VA mycorrhiza in excess Al- conditions has already been proved. Its role is also recorded in coarsed root system species like cassava in low P soils of high humid regions.

Performance of temperate forest tree (*Pinus rigida*) in acid soils in presence of ectomycorrhiza gives a very good response even at very high soil Al concentration. Phosphorus content in these plants maintained a higher P and lower Al – concentrations in the needles even in inadequate Ca and Mg content in presence of ectomycorrhizas.

2.5 Manganese Tolerance

In mature leaves of tolerant species, Mn is uniformly distributed while in sensitive species this does not happen due to accumulation of Mn in a specific site and hence chlorosis and necrosis are quite common. Mn is evenly distributed in leaf blades of both resistant and susceptible cultivars in presence of silicon. In coniferous trees (*Norway spruce*) formation of Mn- complex of high stability with oxalic acid enable it to tolerate in presence of high Mn content in temperate conditions. Tolerance of plant under excess Al- content differs from tolerance of plant to high Mn – content and vice versa. Mn tolerance in vegetative phase may be higher as compared to reproduction (*Glycine max* and *Vigna unguiculata*). Some workers have suggested induction of Mn to the peduncle may be a technique for screening plant cultivars for Mn- tolerance during reproductive phase since this phase is more susceptible to toxicity as compared to vegetative phase. Mn toxicity in acid soils might be the reason for low pods formation in forage cowpea.

2.6 Nutrient Efficiency

In absence of any reclamation to acid soils, the species in question must be efficient in uptake and utilization of P, Ca, B, Mo and Mg nutrients which are generally in deficient conditions. Some of the tropical species like cassava and several tropical grasses (*Brachioria decumbens, Andropogon gayanus* and *Pennisetum* sp.) and legumes (*Stylosanthes* sp., *Macroptillium* sps. and *Cassia* sps.) are quite efficient in uptake and utilization of phosphorus in presence of high concentration of Al in acid soils. This might be due to binding of Al in root apoplasm in presence of high Al. Deficiency of B and Mo is quite common in acid soils of the tropics which restrict the growing of legumes like *Medicago sativa* as compared to cowpea and to some extent clovers. Application of lime and these micronutrients usually gives a good response to forage and seed production.

Chapter 3
Soil Alkalinity

Calcareous soils cover about 30 per cent of the earth surface and mainly presence in semiarid to arid regions. The accumulation of $CaCO_3$ in the upper most layer of alkali soils may be as high as 95 per cent. The pH of these soils is determined by the presence of $CaCO_3$ which buffers the soils in the pH range of 7.5-8.5 sodic/alkali soils are characterized by sodium absorption ratio (SAR)

$$SAR = \frac{Sodium}{\sqrt{Calcium + Magnesium}}$$

It is the relationship between soluble sodium and soluble divalent cations (calcium + magnesium) measured in exchangeable sodium percentage of soil matrix greater than 15 meanings presence of sodium carbonate. Among the sodic and saline- sodic soils, the latter area is much higher than the former. Plant growth on saline soils is affected by high levels of NaCl which impairs the water balance while high pH, high bicarbonate and restricted aeration are responsible for stunted plant growth and toxicity of Na and B.

3.1 The Major Nutritional Stresses on Plant Growth in Alkaline Soils

	Soil pH	
7	8	9
Dominant soil	Calcareous soils	Solontz (sodic soils)
Associations	(Chernozems, Xerosols)	Soloneclaks (saline soils)
Major nutrient	Deficiency: Fe, Zn, P, (Mn)*	Toxicity: Na. B
		Deficiency: Zn, Fe, P (Ca, K, Mg)*
Other constraints	Excess of HCO₃	Poor aeration
	Water deficit	Excess of HCO₃
	Mechanical impendence	water deficit
		Mechanical impendence

*: Less frequently presence

The differences in nutritional constraints (Table 3.1) indicates that toxicity of Na and B, deficiency of Ca, K and Mg are less frequent in sodic soils as compared to calcareous soils while deficiency of Fe, Zn, and P are common in both types and poor aeration is only found in sodic soils

In these soils, nitrogen is bound to humus and becomes available only after mineralization in presence of microorganism which requires optimum soil moisture, temperature and aeration and hence soil pH has the secondary role for availability and harvest of nitrogen in these soils.

3.2 Micronutrients

Iron

Chemistry of Iron in Soils

Biotite, pyroxene, amphibole, olivine, ilmenite, magnetite, pyrite and titanomagnetite are the primary minerals in which iron is a major constituent. These are further grouped into various iron minerals like hematite, goethite, lepidocrocite, ferrihydrite and maghemite under variable time, temperature, oxygen and carbon-dioxide pressure, moisture content, pH, redox-potential, activities of microorganisms and others molecules of *viz.* aluminum, silica, phosphorus and organic molecules. The weathering is also dependent on different layers of the atmosphere; atmosphere, biosphere, hydrosphere and lithosphere.

Iron-Deficiency Stress

On an average Earth's crust contains 41kg/ha of iron which ranges from 2 to 550g/kg of soils. Only 0.1g/kg iron is required for the optimum plant growth but soils pH 7.0 is deficient in iron. Solubility of Fe^{3+} decreases to 1/1000 times/unit increases in solution pH. In alkaline soil solution the concentration of metal ions is also decreased due to formation of precipitates of hydroxides where Fe^{2+} solubility or availability is more affected which occurs in order of

$$Ca^{2+} > Mg^{2+} > Cd^{2+} > Fe^{2+} > Zn^{2+} > Cu^{2+} > Al^{3+} > Fe^{3+}$$

Therefore, occurrence of iron deficiency is more among others in high pH soils. About 25-30 per cent of world's soils are calcareous in the surface horizon leading to iron deficiency In well–aerated high pH soils concentration of both forms (Fe^{2+} and $Fe^{3+)}$ is very low. Application of FYM in calcareous soils low in organic matter may be beneficial in increasing the solubility to facilitate plants uptake at least to low iron requiring crops. Sodic soils of pH > 8.5, increasing the concentration of $NaHCO_3$ from 12 to 75 µM (pH 8.0–8.8) increases the concentration of iron and manganese in the soils solution by 18 and 2.3 times, respectively and thus prevents plants from iron deficiency.

Mechanisms of Iron Acquisition in Plants

Under iron deficient conditions, some plants species have developed efficient strategies for the uptake of iron from low solubility sources. These strategies are of two types.

Strategy I: that adopted by di-cots and non-graminaceous monocots

Strategy II: that adapted by graminaceous

Strategy I

These plants release proton (H^+) and chelators/reductants such as electron (e-) organic acids and phenolics into the rhizosphere and thus increases Fe- solubility and is known as H^+ - promoted dissolution

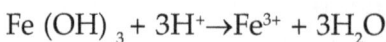

$$Fe\ (OH)_3 + 3H^+ \rightarrow Fe^{3+} + 3H_2O$$

Further secretion of e^- accelerates the dissolution reaction

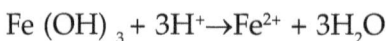

$$Fe\ (OH)_3 + 3H^+ \rightarrow Fe^{2+} + 3H_2O$$

Plants like pea and rape transform crystalline goethite into amorphous iron to acquire the iron but such transforming mechanism is not available in graminaceous plant like maize. Some genes in tomato, rice and other plants have also been identified and isolated which dissolves the unavailable form of Fe^{3+} into available form Fe^{2+}.

Strategy II

Iron acquisition by this strategy of plants is due to secretion of hexadentate Fe^{3+-} chelating substances (phytosiderophores) and by their specific uptake mechanism. Secretion of mugineic acid (MA),which coordinates with Fe^{3+} to form 1:1 complex, dissolve soil iron for plants. Similarly, microorganisms also release Fe – soublizing agents known as **siderophores**. Thus process of Fe- acquition under this strategy by plants can be divided into four main steps.

- ✰ *Biosynthesis of phytosiderophores within the roots.*
- ✰ *Secretion of phytosiderophores to the rhizosphere*
- ✰ *Phytosiderophore- promoting solubilization of low solubility iron by chelation in soils and*
- ✰ *Uptake of phytosiderophore – Fe^{3+} complexes by the roots.*

Genetic Improvement for Iron-Acquisition in Plants

Till date nine MA- related analogues have been identified and isolated from different graminaceous species and cultivars including 2 very important compounds (3- hydroxyl- 2- deoxymugineic acid 2- hydroxylavenic acid A) identified from perennial grasses (*Lalium perenne* and *Poa protensis*).

Introduction of a reconstructed yeast ferric reductase- gene **refre1** into tobacco resulted in enhance tolerance to low Fe- availability conditions in high pH solution. In transgenic heterologous expression of the *A. thaliana* chelate- Fe^{3+} reductase gene, FRO2, significantly increases Fe^{3+} iron reduction in transgenic soybean roots and leaves. It reduces chlorosis and increased chlorophyll concentration as compared to plant under control. A barley genomic DNA fragment containing two neat genes introduced in transgenic rice enhances the root secreation of phytosiderophores under Fe- deficient conditions. Genes introduced rice tolerated the Fe- deficiency and gave more than 4 times higher yield than control conditions in alkaline soils.

Zinc

In high pH soils the solubility of Zn as to that of Fe is reduced since diffusion coefficient of Zn in calcareous soils are 50 times less than acid soils which is dependent on Ca^{+2}, organic matter and microbial activity. Alike to Fe, application of FYM increases the solubility of Zn. Though application of $NaHCO_3$ increases Fe solubility but increases Zn deficiency due to restriction in root growth.

Manganese

Availability of Mn in soil is controlled by redox potential and pH besides these, it forms organic and inorganic complexes. Solubility of Mn decreases with increases in $CaCO_3$ and MnO_2 in aerated calcareous soils due to adsorption on $CaCO_3$ and oxidation on MnO_2 surfaces may be due to formation of Mn calcite. In calcareous soils Mn availability is dependent on the population of anaerobic microbes and root – induced charges in the rhisosphere. Some times Mn- deficiency also occurs due to over liming of acid temperate soils.

Boron

Solubility of B in alkaline soils may be due to adsorption to clay minerals which is usually compensated by boron presence in irrigation water. Boron may be in toxic level in sodic soils and deficiency hardly occurs.

3.3 Major Chemical Constraints in Plant Growth

1. Iron deficiency
2. Zn and Mn deficiency

Iron Deficiency

Lime induced chlorosis due to Fe- deficiency in soils with > 20 per cent $CaCO_3$ is common. Among major cereals; rice usually upland or under water deficient conditions is susceptible to Fe- deficiency. Peanut, soybean in legumes, apple, citrus, grapevines and peaches in fruits are also adversely affected due to Fe-deficiency

In maize and sorghum, iron deficiency is closely correlated with decreasing levels of poorly crystalline or amorphous iron oxides but in other than cereals (apple and soybean) it is correlated with levels of HCO_3^-. Thus the graminaceous plants are termed as strategy I plants and rest as strategy II plants: Root responses to Fe- deficiency are inhibited by high pH which is affected by:

1. impairment of the affectivity of H^+ efflux pump by neutralization of the H^+

2. Lowering the release of phenolics and

3. Fe- reduction at the plasma membrane besides high concentration of HCO_3^-. Sometimes very high dose of P- fertilizer also aggravates the Fe- deficiency in calcareous soils.

3.4 Zinc and Manganese Deficiency

High soil pH together with high clay (per cent), high application of phosphorus under low temperature and water stress in particular are conducive in aggravating Zn- deficiency in cereals. Zn deficiency is also very common in low land flooded rice in high pH soils with high in organic matter. In neutral and alkaline soils, a strong correlation between soil pH and rice yield has been recorded in absence of Zn-nutrient.

Mechanism of Adaptation

On the basis of differences in characteristics the plant species are classified into the three categories.

1. Calcicoles and calcifuges

2. Iron efficiency and Chlorosis resistance

3. Zinc and Manganese efficiency

1. Calcicoles and Calcifuges

The plants that grow on calcareous soils termed as 'Calcicols' due to their adaptive mechanism to sustain under low iron and zinc availability in presence of higher concentrations of calcium and bi-carbonate ions. These species are highly efficient in phosphorus uptake since these are infected with VA mycorrhizas. Some plant species which are adapted in acid soils of higher iron concentration are termed as' Calcifuges'. Their distinguishing characteristics are as under.

Calcicoles

☆ Adapted to calcareous soils

☆ High acquisition capacity for Fe and Zn

☆ Sustain under high Ca and HCO_3^- content in soils

☆ Efficient in P-uptake

Calcifuges

☆ Adapted to acidic soils

☆   Roots exudates contain different oxalic acids that bind the excessive Al and Fe to form complex ions. The entry of such complex ions in xylem vessels is restricted due to increases in size as compared to normal ions.

3.5 Iron Efficiency and Chlorosis Resistant

Plants which are efficient in Fe-acquisition from Fe-deficient soils are known as iron efficient species. These species mobilize iron by both non-specific and specific mechanisms.

a. Nonspecific Mechanism

It is related to Fe-nutritional status of plants. Enhanced root exudation of organic acids occurs under P-deficient acid soils, usually by legumes. In Lucerne, under P-deficient acid soils, root exudation of citric acid increases to almost double. Such exudation is very high in chickpea and peanut, moderate in pigeon pea and low in soybean. In pigeon pea exudation of piscidic acid (p-hydroxybenzyl tartaric acid) in low pH acid soils complexes the ferric iron and releases the phosphorus of ferric-phosphate in available form. That is why pigeon pea is grown in acid soils usually without liming.

Non-specific mechanism of plants mainly occurs due to the following reasons.

   i. Root-induced decrease in rhizosphere pH due to preferences in cation uptake as well as due to application of $(NH_4)_2SO_4$ or N_2-fixation in legumes.

   ii. Exudation of organic acids in P-deficient soils and

   iii. Effect on soil pH, redox-potential and chelators concentration in rhizosphere caused by release of root-photosynthesis as substrate for microorganisms.

b. Specific Mechanisms

Some plant species have two strategies (Strategy I and Strategy II) of root response to iron deficiency.

Strategy I: It occurs only in monocot which is characterized by both increases in reducing capacity as well as enhances in net root excretion of protons. It is also influenced by chelating compounds mainly phenols. These species change both the root anatomy and morphology especially in the formation of rhizodermal cells.

Strategy II: It occurs only in grasses and is characterized by an enhancement in release of non-proteinogenic amino acids (phytosiderophores) such as mugineic acid which forms highly stable complexes with iron and accelerates the transport capacity of this nutrient in roots.

Positive correlation between the extent of enhanced release of phytosiderophores and resistant to Fe-deficiency in calcareous soils occurs in order of,

Wheat > barley > rye, oat >> maize ...sorghum

Release of phytosiderophores is less in flooded rice soils which confirmed that rice being a calcifuges species.

3.6 Zinc and Manganese Efficiency

Acquisition capacity for zinc availability in alkaline soils of species differs which may be due to variations in rhizosphere pH, root colonization and root exudation. As such bean, maize, cotton and apple are more sensitive than wheat, oat and pea. An efficient variety of which requirement for zinc is very high, Zn concentration in leaves may be high even it is grown in Zn-deficient soils and hence hardly any reduction in yield occurs. Though, it is not very clear that enhancement in release of Zn might be due to phytosiderophores biosynthesis by iron and zinc or may be due to a separate iron metabolism under Zn-deficiency. Efficiency in Zn-acquisition also differs among varieties of the same crop. In Zn-efficient varieties, Fe-metabolism is severely restricted under Zn-deficiency rather than in Zn- inefficient varieties.

Zn-efficiency in rice cultivar may be related to high HCO_3^- tolerance of roots and a better control in organic acid accumulation in the roots as compared to Zn-inefficient one. Besides presence of bi-carbonate ions, low root temperature under sub-merged conditions, may have some positive effects on the performance in uptake of Zn, Fe and Mn in high pH soils environment.

Alike to Zn, plant genotypes also differ in Mn –acquisition due to their variations in heredity or genes. Even it varies from one cultivar to another as it is recorded in cases of wheat and barley. Though, mechanism responsible for varietal differences in Mn-acquisition is still not known however, it seems that a particular dynamics of Mn in rhizosphere microorganisms might be responsible for the problems in identification in mechanisms.

Acquisition of Mn in alkaline soils is controlled by presence of Fe. As such, if concentration of iron will be high, the Mn-toxicity will not occur and if the concentration of Mn will be high, the Fe-toxicity will not occur, accordingly.

Formation of special type of root (proteoid) usually in lupin helps in Mn-availability. In calcareous soils P-deficiency increases formation of proteoid root and thus enhances Mn uptake to a toxic level.

3.7 Soil pH, Calcium/Total Cation Ratio

Root growth may be restricted by high pH either directly or indirectly. Functioning of transmembrane pH gradient electro potential gradients and the proton anion cotransport at the plasma membrane are the causes of direct effects on root growth. Application of ammonium phosphate or band application of Urea causes root injury in neutral to alkaline soils. At high soils pH root growth of low land rice and calcifuges species might be restricted also by elevation in HCO_3- concentration.

If pH<5 high activity of monomeric aluminum develops Al toxicity which is restored with application of calcium on an average a molar ratio in the soil solution of calcium total calcium of merely 0.15 is required to optimum root growth. Application of lime in these soils allows deeper root penetration to facilitate xylem to extract calcium. Chelation of Al and mulching is the most effective method, to facilitate root elongation in highly permeable acid soils under saline conditions, application of ammonium phosphate reduces Ca/total cation ratio and hence reduces root growth and also holds good for low pH soils.

Chapter 4
Soil Salinity

Arid and semiarid regions of the tropical 11 countries of south Asia are dominated by saline soils covering 317mha due to deficient rainfall which is inadequate to leach down the salts in lower horizons. Majority of rangeland soils are saline due to lack of irrigation on one hand and excess-evapotranspiration on the other. Intensively canal irrigated areas due to carrying above the salts from lower layer is the serious problem in high intensity crop land covering 43.8mha. In addition to these salt marshes of the temperate world, mangrove swamps of tropical and sub- tropical sea- coast and surrounding lakes are the other problematic area for crop production infected with high salt deposition. About 33 per cent of world irrigated area is affected by salinity of which major parts of Pakistan and border states of India are the worst affected.

Even good quality irrigation water contains enough salt (1to 10/10,000 m^3. ha) which needs leaching and drainage frequently. Under such situation,selection of crop varieties through breeding along with some agronomical manipulation in planting patterns may solve the problem to some extent.

4.1 Characteristics and Types of saline Soils

The US Salinity Laboratory defines that saturation extract (solution extracted from a soil at its saturated water content) of a saline soil has an EC > 4mmho/cm or 4deci Siemens/cm (equivalent nearer to 40mM NaCl/

L) and ESP < 15. ESP > 15 is termed as saline-alkaline soils or saline-sodic soils has high pH values and tends to restrict aeration and water movement. A wide $Na^+ : Ca^{2+}$ ratios in the substrate is the distinguishing character typical for sodic soils but not for saline soils.

EC of a good quality water should be < $2dSm^{-1}$ while EC of sea water may be as high as $55dSm^{-1}$ or almost 28 times higher to irrigation water. Since salt concentration in the root zone soils may be much higher than the upper surface and EC only gives the concentration of salt but not their composition however NaCl is the dominant salt which governs the intensity of salinity.

4.2 Salinity and Plant Growth

Eco-physiologically the résistance to salinity of plant types are in the order of,

Halophytes > Halophytic crop species (sugar beet) > salt tolerant crop species (Barley, Bermuda grass and Rhodes grass) > sensitive crop species (bean).

The threshold EC and the percentage of yield decrease beyond threshold for the salt tolerance/sensitivity of crop species of food, forage and fruit crops can be an ideal guide for managing crops under the intensity of salinity (Table 4.1).

Table 4.1: Tolerance of Crop Species to Soil Salinity. Threshold EC_e (25⁰C) = Maximum soil salinity that does not reduce yield; slop= yield reduction/unit increase in EC beyond threshold (Mass, 1985).

Crop Species	EC Saturation Threshold (dSm⁻¹)	Soil Extract Slop (per cent/dSm⁻¹)
Barley *(H. vulgare)*	8.0	5.0
Sugarbeet *(B. vulgaris)*	7.0	5.9
Bermuda grass (*C. dactylon*)	6.9	6.4
Wheat (*T. aestivum*)	6.0	7.1
Soybean (*G. max*)	5.0	20.0
Tomato *(S. esculentum)*	2.5	9.9
Maize (Z. mays)	1.7	12.9
Orange (*C. sinensis*)	1.7	16.0
Grape (*Vitis* sps.)	1.5	22.0
Bean (*P. vulgaris*)	1.0	19.0

Besides genetical variability of the crop species, even different cultivar of the same crop varies in their salt tolerance capacity. As such some significant progress has been made in cases of sub- merged rice and barley Stages of the crop growth period are also differently affected by salt. As such sugar beet at germination while rice, wheat, tomato and barley after germination and maize at tasseling and grain feeling stages are more susceptible to salt toxicity.

4.3 Plants Adaptation Mechanisms to Salinity

Similar to plants adaptation mechanisms under adverse effects of acid and alkaline soils environments, Greenway and Munns (1980) have also classified the plants as excluder and includer (Figure 4.1).

Plants may adapt either of the two above mechanisms for their survival under saline soil conditions. Salt exclusion needs mechanisms for avoidance of an internal water deficit while adaptation by salt inclusion needs high tissue tolerance to both Na^+ and Cl^- or avoidance of high tissue concentrations.

High salt tolerance in halophytes of the *Chenopodiaceae* (sugarbeet) is based on the inclusion of salts for their utilization in maintaining turgidity of replacement of K^+ in different metabolic functions by Na^+. Among excluders, *Casurarina* sps, *Puccinellia peisonis* and *Festua rubra* are the most salt tolerant plants.

Among cultivated cereals, maize is less tolerant to salinity than sugarbeet due to the fact that even at low level of Na^+ and Cl^- in its shoot, growths restricted due to obstruction in osmo-regulation but not due to deficiency of K^+ and Ca^{2+}. Among legumes, bean is the most sensitive to Cl^- toxicity but not of Na^+ as such however, soybean is capable to exclude Cl^- which is controlled by a single gene pair.

Salt sensitive species, *Phaseolus vulgaris* and salt tolerance species, *Trifolium alexandrinum* and *Phragmites communis* sustain under saline soil conditions due to re-translocation of Na^+ from shoots to roots. Presence of well developed **Casparian band** in the halophytes, mangrove (*Avicennia marina*), acts as a barrier to restrict passive influx of salts in this is the main mechanism to survive under sea water.

In barley and sorghum roots enhancement of plasma membrane and tonoplast-bound H^+ pumping ATPase activity may be responsible for their salt tolerance. Introduction of **D-genome** from the salt tolerant diploid *Agilopes squarrosa* and producing hexaploide type makes it possible to release salt tolerant varieties of wheat through breeding since D-genome increases the K^+/Na^+ selectivity in both roots and shoots of hexaploid plants. However, in barley high salt tolerance is due to higher xylem transport of Na^+ to the shoots, is a best example of salt encluder. In tomato, restricted translocation of both Na^+ and Cl^- to the shoots while in some other, accumulation of salt in the shoots is responsible for proper growth.

Shifting in C_3 photosynthesis to CAM facultative halophytes *(Mesembryanthemum cryatallinum)* decreases transpiration and thus restricts the salt transport to the shoots and protects the plant from salt injury.

Salt excretion is the other mechanism to avoid salt injury. *Avicennia marina* halophytes accumulates salts in bladder hair, secrets salts glands,

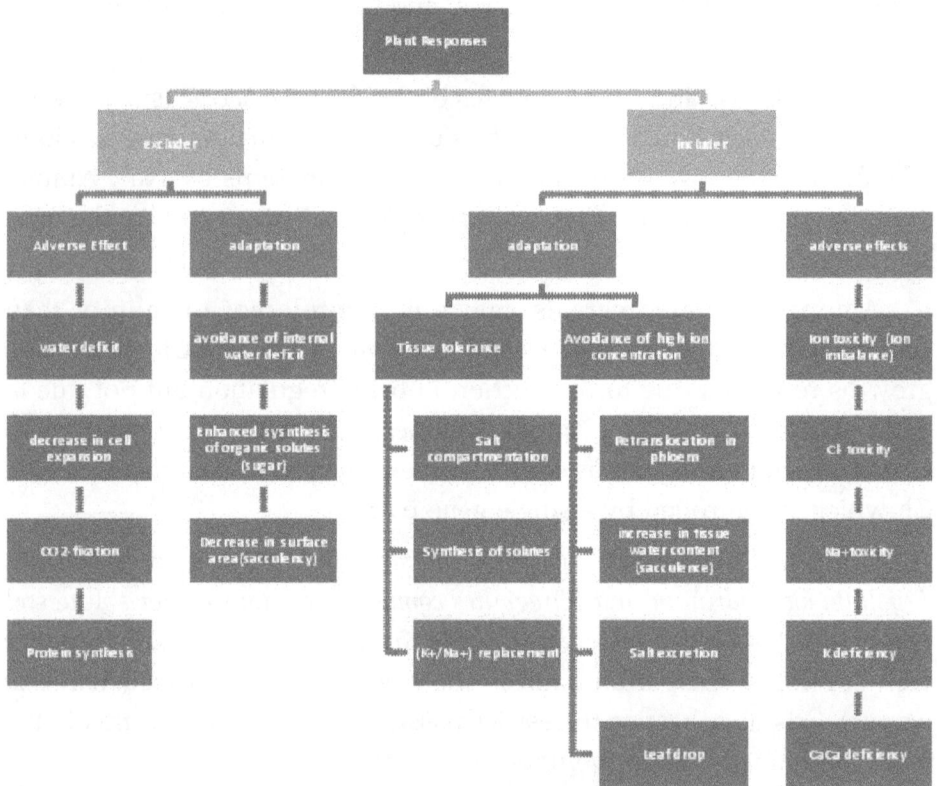

Figure 4.1: Adverse Effects of Salinity and Adaptation Mechanisms (Greenway and Munns, 1980).

sheds older leaves and retranslocates the salts to other organs to sustain in saline environment. Salt glands are also present in *Leptochloa fusca*, secrets salt on the leaf surface which washes out at raining or falling dew, is a temperature dependent process.

4.4 Major constraints of Soil Salinity

Water deficit, iron toxicity and nutrient imbalance by depression in uptake and transport within the plant are the major constraints of salinity for optimum plant performance.

4.4.1 Water Deficit and Salinity

Adverse effects of salinity on shoot growth are more than root development as such leaf elongation rate is reduced due to decrease in leaf water content as it is noticed in bean, wheat and barley plants. Substrate salinization decreases both water availability and uptake which results in reduction of pressure-driven xylem transport of water solutes (salts) and therefore, water as well as nutrient status of shoots are reduced.

Since the saline substrate is mainly consisted of Na^+ and Cl^- ions and even Cl^- being an essential nutrient for plants and Na^+ an essential mineral for a number of halophytes and C_4 plants but extreme concentration of the same results toxic for them. Many fruit plants *viz.*: grape vines and citrus and avenue trees *viz.*: *Aescules* and *Tilia* are sensitive to high Cl^- ions which causes chlorosis. Hence, application of KCl fertilizers in coastal soils further aggravates the Cl^- toxicity.

In rice, wheat, sorghum and other cereals Cl^- toxicity is more accountable than Na^+ toxicity under Ca- deficient conditions. Salt accumulation in the leaves apoplasm causes toxicity resulting in dehydration and loss in turgor pressure and hence death of leaf tissues and cells. Besides these uptake transport and utilization of nutrients may be depressed particularly of Mn^{2+} while Cl^- concentration may restrict NO_3^- uptake.

In case of P nutrition, NaCl concentration may increase P availability and even P toxicity but contrast to this, under high NaCl level and low P level the reduction in the yields of cotton and tomato have been recorded due to depression in P utilization efficiency. In spinach plants, K demand may be doubled for the same quantity of photosynthetes needed for photosynthesis as compared to normal conditions.

4.4.2 Role of Calcium in Salt Affected Soils

Application of gympsum in saline-sodic and sodic soils is a common practice, which signified the role of calcium (Table 4.2). A marked (149 to 207) increase in potato tubers is recorded even at high salt content (1.2 per cent). The application of gypsum also improves soil structure, aeration and Ca^{2+}/Na^+ ratio and thus restricts Na^+ entry in to the roots.

Table 4.2: Effect of Salinization and Gypsum on Potato in Sandy Loam Soil.

Treatment	Tuber Yield (g fresh wt./plant)	
	Without Gypsum	*With Gypsum (2 per cent)*
Control	221	226
0.6 per cent salt	183	280
1.2 per cent salt	149	207

Salinization (Na^+, Mg and Ca salts), Abdullah and Ahmad, 1982).

Application of calcium in bean, citrus and rice results in drastic reduction in NaCl concentration in leaves due to development in selectivity of ion uptake and transport in presence of calcium. Increasing calcium concentration from 0.1 to 4,8 mM in soils increases the Ca content in shoot dry weight in rice from 0.07 to 0.14 per cent. Beneficial effects of gypsum on *Brassica* sps., barley and wheat have also been recorded.

In horticultural plants, Ca-related physiological disorders like tip burn in lettuce and blossom-end rot in tomato are very common but in Chinese cabbage (*B. pekinensis*) increase in salinity may not affect green weight adversely even though a marked incidences of tip burn.

Salinity-induced Ca-deficiency in terms of Ca^{2+} homeotasis at cellular level in the roots and transmitting a signal of salinity stress in roots and shoots has been reported recently. High Na^+ concentration displaces Ca^{2+} from the sites on the outer surface of plasma membrane of the root cells or more likely from intracellular membranes and thus impairs Ca^{2+} homoeostasis in cells and its role as secondary messenger.

Plants grown under saline soil conditions usually have low photosynthesis efficiency but have high respiration and thus low net gain as well as low in leaf protein content. Foliar application of ABA increases CO_2 fixation and supply of CYT photohormone reduces toxicity of NaCl in plants.

4.4.3 Genetic Variability in Salt Tolerance

Some for the plant species and their cultivars of different cereals, legumes, oilseeds, vegetables, fruits, forages and ornamental plants have been identified for growing in saline soils are listed (Table 4.3).

Table 4.3: Crops and Varieties Screened for Growing in Saline Soils of Respective Country.

Crops	Variety	Country
Cereals		
Wheat	SAR C- 1, Lu- 20 S, Karanchi- Rata	Pakistan
	Sakha- 8	Egypt
Sorghum	Desert maize Shallu, Hegari	Pakistan
Maize	Arizona 8601	Egypt
Rice	Giza 159	Egypt
Field Crops		
Cotton	Bahtima 110' Menoufi,	Egyptian
	Ashmouni and Bahtim 108	
	Cotton varieties	
Sugarbeet	U.S.H 9, Maribo Magnapoly, and kawep	India
Oil seed crops		
Rape seed, Canola	(B. sps.) a derivative of rape seed	
Sunflower	HO-1, Predovik and Euroflor	Pakistan

Grasses and Forages

Cynodon ductylon, Andropogon desertorum, Poa pratensis, Lolium perenne, Dactylis glomerata, Agrostis stolonifera, Chloxis gayana, Festuea, rubra.

Some of the legumes like *Leucaena leucocephala* and *Centrosema pubescence* are very resistant to salinity while *Medic* sps., and clovers (*Trifolium, Repens* and *I. subterraneum*) are moderately tolerant to salinity.

Vegetable Crops

Most of the vegetable crops are sensitive to salinity including changes in taste and quality. However, flavour of carrots and asparagus is increased but yield decreased.

In muskmelon, BG 84 -3 of Israel and Shad EL Dokki of Egypt have been tested for their salinity tolerance. In lettuce (*Lactuca sativa*), Romaine varieties are more resistant to salinity than iceberg varieties.

Fruits

Most of the fruit plants are sensitive to salinity and Cl⁻ is more injurious than Na^+. Among fruits, date palm (*Phoenix dactylifera*) and pomegranate (*Punica granatum*) are moderately tolerant.

Ornamentals

Application of tissue culture technique has not achieved any success as yet however, some attempts are being made in lucerne and bent grass. Similarly, molecular biology also has not achieved any breakthrough in this respect.

Chapter 5
Flooded Soils

Waterlogged and flooded soils are used synonymously to refer soils with excessive water levels. Flooded or submerged soils are permanently below the water table for months during the year under such situation, oxygen is depleted due to respiration of soil microorganisms. After the utilization of molecular oxygen in respiration different microorganisms use to take terminal electron acceptors for respiration and hence a sequence of reduction takes place at specific redox- potentials which varies as per depth and series of reactions usually occurs.

When free oxygen is depleted NO_3–N is used by the microorganisms

NO_3^- reduces to NO_2^- to N_2O to No and finally N_2 under dry and wet conditions, NO_3^- can temporarily accumulate.

Under acid soils conditions, heterotrophic nitrification that is nitrite and different nitrous oxides are formed ($NH_4^+ \rightarrow NO_3^-$) under aerobic conditions.

Acid soils high in manganese oxides and organic matter but low in NO_3^- changes manganese into soluble form and causes Mn toxicity.

Water logging for longer period reduces ferric form of iron to ferrous form and increases the availability of iron- phosphate (Fe-P).

Reduction of sulphate to H_2S decreases the solubility of iron, Zinc, Copper, Calcium and Cadmium by forming soluble sulfides.

Formation of sparingly soluble Zn- compounds may create Zn - deficiency in submerged rice.

Besides changes in nutrition of essential mineral elements, formation of ethylene occurs due to decomposition of OM. Low redox potential activates CH_4 formation in submerged rice and releases into the atmosphere to enhance pollution.

5.1 Plant Performance

Decrease in the hydraulic conductivity of roots leading to ethylene accumulation in roots results in their wilting. Adverse effect of water logging depends in plant species, soil physical and chemical conditions. As such rice and para grass (*Brachiaria mutica*) are more resistant to submerging as compared to legumes (Peas). High temperature is further detrimental for aerobic plants like wheat and barley.

Severe reduction in uptake and translocation of nutrients due to cessation of root growth and respiration occur within few days of flooding. Lack in root aeration leads to decreases in concentrations of the three major plant nutrients (N, P and K) in several cereals (Maize, wheat etc.), Continuous water logging causes manganese toxicity especially to Lucerne grown in acid soils.

Even prolonged water logging in low pH soils high in iron, causes 'Bronzing 'of leaves in rice due high concentration of iron (> 700 mg Fe/ kg on dry weight basis) since under high iron content, activity of polyphenol oxidases is increased in leaves resulting "bronzing" as well as "brown speckles" due to Mn^{2+} toxicity. Low concentration of oxygen within root zone enhances the selectivity for Na^+ in place of K^+. Increase in temperature under arid soil conditions further increases the Na^+ uptake by plants (sunflower and apple trees). Since high salinity and aeration are injurious for plant hence due attention is needed while irrigating poorly structured soils with saline water.

Acid soils high in OM, prolonged water logging may accumulate volatile fatty acids phenolics in the soils affecting root metabolism and growth. Continuous submergence also produces ethanol which enters into the xylem and damages the plant cells. Excessive formation of ethanol in apple plantation reduces the keeping quality of fruits. Activity of ethylene production in soils poor in N may be very high.

5.2 Mechanisms of Adaptation

Plant may be adapted to water logging (O_2 deficit) either by avoidance of stress or tolerance of stress or both as it varies from species to species

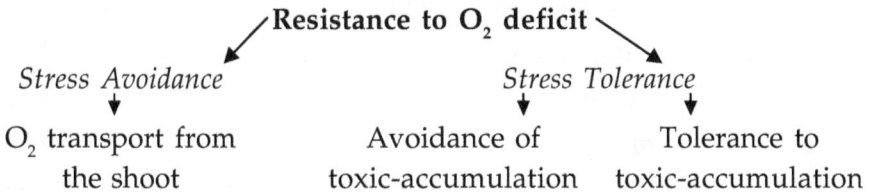

Resistance to O_2 deficit

Stress Avoidance   *Stress Tolerance*

O_2 transport from   Avoidance of   Tolerance to
the shoot   toxic-accumulation   toxic-accumulation

Presence or absence of O_2 is the prime factor for both avoidance and tolerance. Tolerance may be usually temporary but avoidance adaptation may be under permanently submergence. Sustaining under high Fe $^{2+}$, wetland species requires special mechanisms to detoxify the iron either by exclusion from uptake or uptake (includer) and detoxification by SOD (Superoxide Dismutage) an antioxidant, peroxidase and catalase.

$$Fe^{2+}O_2^- \xrightarrow{SOD} H_2O_2 ---Peroxidase--- H_2O$$

5.3 Metabolic Adaptation

Flooding – tolerant plants are better equipped to regulate their rate of glycolysis and fermentation to ethanol in comparison to flooding- sensitive plants. As such in barley under hypoxia the activity of alcohol dehydrogenase (ADH) increases to 10 times while cytochrome oxidase activity decreases significantly. Since only rice is able to maintain energy charge in presence of high carbon flow to ethanol hence it is resistant to water logging.

Flood resistant species increases SOD activity under anoxia and prevents oxidative damage. Since under water logged conditions, concentration of O_2 in the root tissues are lower than the soil solution hence roots already experience hypoxia before the onset of anoxia.

Under submerged conditions, Mn-toxicity is generally dominated but complexing of Mn with organic acid, reduces the Mn-toxicity. As compared to upland rice genotypes, lowland genotypes have high tolerance to iron toxicity up to having 1100-1600 mg Fe/kg leaf dry weight. Sometimes, transformation of special types of anaerobic polypeptides (protein), are formed by the resistant genotypes which protect the plants from adverse effects of submerging.

Anatomical and Morphological Adaptation

Variations in root anatomy and morphology of different plant species are also responsible for their adaptability under flooded soils conditions.

☆ Plants with higher root porosity have high degree to sustain under water logged conditions as it happens in lowland rice cultivars.

☆ Flood tolerant plants develop an extended aerenchyma both in roots and rhizomes and thus facilitate better oxygen movement.

☆ In *Rumex* species, rice and sedges formation of aerenchyma is a continuous process and is enhanced by flooding.

☆ Changes in root anatomy with changes in morphology also occur, such as development of aerenchyma is enhanced in the newly developed roots.

☆ This aerenchyma in the basal zones occupy up to 50 per cent of the total root volume in flooded soils as compared to only 20 per cent in drained soils.

☆ Under flooded conditions, when majority of the rice leaves are submerged in the early part of growth, the main path of oxygen diffusion in the shoot tissue appears to be along a continuous air strip between the hydrophobic leaf surface and the surrounding water rather than in the leaf aerenchyma.

☆ Release of molecular oxygen by the roots creates oxygenation of the rhizosphere to help the rice plant for growing under anaerobic conditions. Such oxygen also reduces the toxic effects of high iron due to changing of ferrous to ferric form.

☆ Under submerged conditions, the increases in silican content in rice plants, the iron uptake may go down to one third.

☆ Mangrove swamps in permanently waterlogged soils avoid iron and H_2S toxicity due to its internal ventilation and presence of tannins which oxidize both Fe^{2+} and H_2S to sparingly soluble FeS.

Mineral Nutrition of Plants under Flood Stress

Loss in aeration or oxygen has a vital influence on nutrient acquisition by plant which is controlled by soil pH and their redox potential.

Oxygen and Soil pH

Since anaerobic conditions favour de-nitrification of the soil nitrate to nitrogen gas and H_2S and hence rise in soil pH occurs but severe leaching of bases and increased CO_2 in flooded soils decreases soil pH. Extent of these two conditions results in variation of pH.

Oxidation-Reduction Potential

Redox potential (Eh) of a soil depends on presence of electron acceptors (O_2 and other oxidizing agents) and favourable pH. In an oxidized soil, Eh ranges from c + 600 to350 mV and in reduced soils or under anaerobic soils from c – 300 to + 350mV. Soil rich in OM with warm conditions, the Eh may be as low as – 300mV.

Nutrient Availability under Flooded Conditions

Reduced aeration under flooded conditions, the uptake rates of N, P, K, Ca and Mg is reduced since ATP supplied by anaerobic respiration is inadequate to supply sufficient energy required for nutrient uptake.

Lack of sufficient N and other nutrients, normal leaf growth and leaf elongation are restricted by root hypoxia for a short period. Inadequate oxygen also restricts the acquisition and transport of nutrients as well as on Eh and pH which are responsible for nutrient availability. General reduction in NO^{3-}, SO_4^{2-}, Mn^{4+} and Fe^{3+} occurs where nitrate is reduced to nitrite and finally to nitrogen gas to escape into the atmosphere. Prolong flooded conditions SO_4^{2-} is reduced to H_2S gas. Solubility of Fe^{3+} and Mn^{4+} increases with anoxia since these are reduced to Fe^{2+} and Mn^{2+}, respectively are toxic in acid soils of the humid regions. A change in soil pH increases the availability of P and Mo but of Zn is reduced due to increase in soil pH. Availability of Cu decreases with increase in pH.

Transport of Nutrients

Flooded conditions inhibit the root transport of nutrients due to decrease in permeability and growth of roots. Under anaerobic conditions, phloem unloading ceases and thus transport of metabolites and growth regulators from root to shoot is stopped. Reduction in water and nutrients transport through xylem to shoot is also impeded due to increase in hydraulic resistance of the root.

5.6 Soil Microorganisms

Soil air is the main source of oxygen for most of the soil microbes – bacteria, fungi, actinomycetes and algae to produce CO_2 in absence of gaseous exchange. CO_2 accumulates in the soil air and becomes toxic to the microbes. Under restricted oxygen conditions, anaerobic organisms predominate and use. Presence of NO_3^-, SO_4^{2-}, and Fe^{3+} as electron acceptors in place to oxygen limits nutrient availability. Free-living and symbiotic N-fixing bacteria functions are also reduced in absence of adequate oxygen but activities of denitrifying bacteria are significantly enhanced.

Continous flooding invites many soil-borne diseases in plants. In tomato, colonization of roots by *Pythium* spp. is reduced while bean is fully infested with *Fusarium solani*. Low O_2 diffusion rate results in reduction of growth and yield. Plant spp. having low glyceoline content are more resistant to disease under waterlogged conditions.

Anaerobic conditions inhibit the synthesis of indole acetic acid (IAA),gibberalic acid (GA) and cytokinins by roots while production of abscisic acid(ABA) is accelerated.

Crop Tolerance/Avoidance of Water Logging

Significant variations in susceptibility to water logging among plant species are recorded. It may be extended from only a few hours to several days depending also on the growth stages.

Morphological adaptation is based on development of arenchyma and internal aeration pathways. In general plants with hollow stems having access to oxygen are more resistant to water logging.

5.7 Flood Problem and Solution

Water deficiency and flooded conditions are equally damaging for crop performance. Flood damage valued 1 to 2$ billion in US itself. Changes in rainfall pattern due to global warming resulted in occurrence in flood in traditionally drought prone area of western India while drought like situations in high rainfall rice bowel of East and North-East parts of India since last decades has led to plan to change in crop and cropping systems in these states. Rivers originating from neighboring Nepal have virtually become sorrow for north Bihar and some parts of Uttar Pradesh which needs an international planning and agreement between the two countries. However, there may be several other reasons of floods.

5.8.1 Causes of Floods

Extreme exploitation of natural resource with in the name of development without forecasting the consequences is the prime cause of flood.It is said, if you abuse the nature it will take its own action. The very recent 'Himalayan Tsunami" in Uttarakhand state (India) in which several thousands of human life perished is one of the classical example of unplanned construction of roads, houses and lakes led to unprecedented landslides and flood. Chronic flood in North Bihar every year is the another example of ineffective planning. In addition the other causes may be as under:

☆ Loss in vegetative cover in the catchments area of the rivers

☆ Over grazing

☆ Faulty planning of urban development; highways, airport,industries and other constructions

☆ Encroachment in swamp area along the rivers banks

☆ Too much rain in short span

5.9 Flood Control

Flood is equally or even more damaging than drought, leading to loss in life and vegetation which is a challenging task for several countries. Some of the measures that can control the misery of floods are:

a. Connecting the rivers as a national perspective

b. Construction of dams and levees along the banks

c. Streambed channelization- a kind of streamlining of rivers and streams, carried out by cutting or bulldozing vegetation along the stream's banks and deepening and strengthening river channels

d. Creating more and more watersheds with sufficient number of lakes and ponds

e. Replanting on the banks and restoring wetlands.

5.10. Salt Stress

As per salt requirements, the plants are classified in to two categories.

1. Glycophytes (Sensitive to relatively high salt concentrations)
2. Halophytes (Adapted to high salt concentrations, usually such are not plants)

The different properties of the Sea water and Irrigation water give indication of good quality water required by a normal plant.

Table 5.2: Properties of Sea Water and of Good Quality Irrigation Water.

Property Concentration of ions (ppm)	Sea water	Irrigation water
Na^+	457	< 2.0
K^+	97	< 1.0
Ca^{+2}	10	0.5 – 2.5
Mg^{+2}	56	0.25 – 1.0
Cl^-	536	< 2.0
SO_4^{-2}	28	0.25 – 2.5
HCO_3^-	2.3	< 1.5
Osmotic potential (MPa)	– 2.4	– 0.039
Total dissolved salts (ppm)	32,000	500

The euhalophytes (true halophytes that tolerate or endure high levels of salts) can grow well where salt levels in the soils are as high as in desert or in soils saturated with brackish waters on the sea coasts where salt contents may be as high as 26 per cent by weight. Species like Iodine bush (Allenrolfea), Pickle weed or samphire (Salicornia), sea lavender (Limonium) and marsh rosemary are the most adaptive to high salt content. Species of *Atriplex* (shadescale) and *Sacrobatus* (black greasewood) are also tolerant to some less salty soils. Genus *Halobacterium* (prokaryotes) accumulate large amounts of salt in to their cells and cannot survive except salty environment.

The halophytes like *Atriplex triangularies* is known as salt accumulator in which osmotic potential continues to become more negative throughout the growing season as salt is absorbed to the extent that it contains 10 – 100 times more salt as is actually observed. At this stage even the water moves into the plant osmotically and not simply in bulk flow.

Some of the cultivars of wheat and mangroves limit the accumulation of Na^+ and Cl^- ions by 100 per cent of the salt by roots exclusion since it contains proline compound. Some plants like greasewood (*Sarcobatus vermiculatus*) bring salts from depths, depositing them on the surface to absorbed atmospheric moisture.

Similar to Na^+, enough K^+ ions creates problem since Na^+ ions compete with the uptake of K^+ by a low affinity mechanism and K^+ is usually present in such soils in much lower concentration than is Na^+. However, if sufficient Ca^{3+} is present, a high affinity uptake system having affinity for K^+ transport can operate well and plant can grow well by restricting Na^+ entry hence gypsum application is preferred in these types of soils.

Chapter 6
Drought Stress

Between soil and plant, it is the water that maintains the turgidity within the plan cells and takes the nutrients from soil in solution form to supply the plant and therefore plant life cannot sustain without water. Since water shares for 80-90 per cent of the fresh weight of most of the herbaceous plant structure and more than 50 per cent fresh weight of woody species, so the importance of water to plants is oblivious.

Land survey revealed that 36 per cent of earth area is under semi-arid conditions receiving only 125 to 750 mm of rainfall annually and remaining 64 per cent is exposed to temporary drought during crop period, about 33 per cent of potential arable land suffers from inadequate water supply and hence a marked reduction in yields occurred. Water is readily lost from the soils, particularly those having a poor water retaining capacity even if, the area is receiving more than 1200 mm of rain fall.

Water is continuously lost from a fully" **saturated soil**" (- 10 kPa), by free draining, under the influence of gravity, and the rate of loss slow down until no water drains away is termed as "**field capacity**" (below – 1500 k{Pa). Further loss of water is due to evapotranspiration and the last stage at which water losses from all these phenomenon is stopped, is termed as the '**wilting point**" (below -3100 kPa) at which water uptake is completely restricted to meet the demand and plant wilts and dies from moisture stress, at the "**permanent wilting point**" (above – 3100 k Pa) at which water remained in the micro-pores as hygroscopic water or unavailable water.

Water stress can affect the plants in the form of "physiological window" mild stress induces in plant regulation of water loss and uptake allowing maintenance of leaf relative water content (RWC) within the limits where photosynthetic capacity and quantum yields show little or no change. The most severe form of water deficit is desiccation – when most of the protoplasmic water is lost and only a small amount of tightly bond water remains in the cell. On the basis of differences in nature of the phytosynthetic apparatus during desiccation two groups of desiccation tolerant plants *viz.*: homochlorophyllous desiccation tolerant (HDT) and poikilochlochlorophyllous desiccation tolerant (PDT) are identified.

Moisture stress primarily results in stomata closure, reduced in rate of transpiration and loss in water potential and hence decreases in photosynthesis, growth inhibition and other physiological processes, even changes in morphology also occurs

Globally agriculture and industry are the major consumers of water (92 per cent). Major volume of water comes from surface water supplies - rivers, streams and lakes. Virtually almost all the nations experiences shortage of water which is predicted to become worse by 2025. The shortage is estimated to be 2.6-3.1 billion compared to 434 million of present day. In several less developed countries together with in arid to semiarid countries do not have facility to clean or even sufficient water for drinking. Unfortunately, rainfall is not evenly distributed and these regions annually receive 250 cm or more rain while only 25cm in desert. According to World Watch Institute, between1950-2050 there will be 74 per cent fall in available water for each person.

Generally, if the rainfall is less than 70 per cent continuously for 21 days or longer, it is termed as drought conditions. Drought also causes loss in ground water resulting unavailability in drinking water and loss in forest area.

Restoration of watersheds and wetlands for mitigating water shortage needs due attention on priority basis in these areas vis-a-vis serious planning to control the population growth are the needs of the hour while, protection of the environment through restoration of vegetation in watershed- replanting forest and grasslands to reduce sedimentation in streams and reservoirs may be the priority area to recycle the water effectively.

The global warming has changed the rainfall pattern of the world which has resulted in an increase in semi- arid to arid and semi-arid to desert conditions due to decreases in total rainfall on one side and increases in flood and marsh lands on the other. These two extreme conditions need special agriculture attention.

6.1 Drought/Desert Conditions

Stress is an external factor that exerts a disadvantageous influence on plants. Some environmental stresses like air temperature may become stressful in just a few minutes while soil moisture content, may take days to weeks and soil nutrients deficiencies be able to take months to become stressful. Water and salinity are the two common stresses which usually occur under drought or arid to semiarid environments

6.1.1 Causes of Water Stress

Water stress causes several disorders in plant anatomical, morphological characters and even in gene expression.

    i. Decrease in leaf area and stimulates leaf abscission

    ii. Short internodes formation

    iii. Deeper root formation in search of water

    iv. Closing of stomata in response to abscinic acid

    v. Limits in physiology within the chloroplast

    vi. Increases resistance to lipid-phase water flow

    vii. Increases in wax deposition on the leaf blades

    viii. Alters energy dissipation from leaves and

    ix. Osmotic stress may change gene expression

Plant species differ in their capacity to sustain under stress conditions. Plants growing under stress conditions usually refer to **stress tolerance or stress resistance.** If tolerance increases due to prior stress, the plant is known as **acclimated** one. However, acclimation can be distinguished from adaptation, which generally refers to a genetically determined level of tolerance acquired by a process of selection over several generations.

Severe water deficit and drought conditions in plants may develop different resistance mechanisms:

i. Desiccation postponement (the ability to maintain tissue hydration)

ii. Desiccation tolerance (the ability to function while dehydrated)/ drought tolerance)

iii. Drought escape/avoiders that complete their life cycles during the wet season or before the onset of monsoon.

Table 6.1: Classification of Plants as per their Water Requirement.

Sl.No.	Types	Character
1.	Hydrophytes	Water loving plants of lakes, ponds and marshy lands
2.	Mesophytes	Water requirements intermediate includes all the major cultivated plants
3.	Xerophytes	Grow under scarce water conditions

Desert plants are termed as xerophytes. These may be known as escape, resist, avoid and endure. Deep rooted plants such as *Prosopis glandutosa* and *Medicago sativa* which do not experience negative water potentials are known as water **spenders.** The other type is called as **desert ephemeral**, is the annual that escapes the drought by existing only as dormant seeds during the dry season and as the rain occurs, these seeds germinate and grow to maturity to set at least one seed per plant before the loss of moisture.

Cacti and Century plant (*Agave americana*) and other CAM plants are water collectors; they resist the drought by storing water in their succulent tissues and loss of water is also very low due to thick cuticle as guard cells are supported by double layers of cells as well as stomata remained closed during the day time.

Several non-succulent desert species have other adaptation that reduce water loss, termed as water savers due to their small leaf blades, sunken stomata, shedding of leaves during dry periods and thick pubescence on the leaves and stems.

Accumulation of various organic compounds, such as sucrose, amino acids (prolin) and several that lower the osmotic potential and thus the water potential in cells without limiting the functions of enzymes that also help in loss of water from plants.

The true xerophytes such as creosote bush (*Larrea divaricota*), a desert shrub is known as **euxerophytes,** drops 30 per cent of the final fresh weight before the leaves die. Some of the most spectacular **euxerophytes** are called as **poikilohydric**, are the mosses and ferns, which are adapted to wet environments but are also suited under very dry conditions due to their ability to reabsorb moisture after drought conditions are also fit for desert conditions. *Selaginella lepidophylla*, *Polypodium* (a fern) and some C4 grasses fall under this group.

Some algae, lichens and mosses can absorb water directly from dew, rain or even a moist atmosphere to continue their metabolic activities. Identification of plants with heteroblasty character or the ability to produce morphologically and physiologically different seeds with different germination requirement, hence only a few seeds in a given crop may germinate at any given time. This covers the risk in seedling survival under variable environmental conditions and some times over several years.

Some plants with characteristics to open stomata when water stress is low and closed when water stress and temperature are high. Secretion of gum and salt on the leaf blades are the other physiological mechanisms which reduces losses of water through leaves. The deposition of salts on leaves also absorbs the atmospheric moisture and then absorb into the leaves. Even the leaves of some of the plant like *Leucaena leucocephala* performs xeric movement (covering of the leaves blades by one another) also reduce loss of water from plants. Such plants also perform di-photonic movement (movement of the plant roots towards soil moisture) also facilitates the plant to absorb moisture from deeper soil strata.

6.1.2 Sensitive to Water Stress

In plants cellular growth is the most sensitive to water stress among the different growth stages at 0.1MPa, the growth of roots and shoots decreases and under this conditions the growth occurs mainly at night. The inhibition of cell expansion is closely followed by a reduction in cell-wall and as well as in protein synthesis.

Activities of many enzymes, especially NO_3-reductase, phenylalanine ammonia lyase (PAL) and few others, decrease very sharply as the water stress increases. N-fixation and reduction also decrease with water stress. As the stomata begin to close, the transpiration and photosynthesis also start to decrease, accordingly.

Resistant cultivars have higher levels of ABA when they are exposed to water stress. ABA concentration in leaves increases in response to several kinds of stress, including nutrient deficiency or toxicity, salinity, chilling and even water logging as ABA is a kind of universal stress hormone, its production is controlled or triggered by several mechanisms.

At higher stress (1.0 – 2.0 MPa) respiration, translocation of assimilates, CO_2-fixation drop to level nearer to zero. Generally plants recover if, watered when stress are -1.0 – 2.0 MPa meaning that, inspite of the sever water stress the plant recovers by shedding of older leaves. Since growth is very sensitive to water stress, yield can be markedly reduced even in situation at moderate stress.

6.3 Survival Mechanisms of Plants at Water Stress Conditions

Physiological and ecological strategies that plants sustain to cope with water stress either by avoidance or tolerance may be classified as;

Short term changes related to mainly physiological responses (linked to stomatal regulation) Acclimation to availability of certain level of water (solute accumulation resulted with adjustment of osmotic potential and morphological changes).

Adaptation of water stress conditions (sophisticated physiological mechanisms and specifically modification in anatomy)

Several processes affect the fitness of a plant in water stress conditions. The adaptations by which plants survive in regions subject to drought, in addition to drought tolerance is known as "drought avoidance". Some of the strategies of drought tolerance may be as

- ☆ Rapid maturation before onset of drought or reproduction after rain
- ☆ Postponement of dehydration by developing deep roots
- ☆ Protection against transpiration or storing water in new tissues and

☆ Allowing dehydration of the tissues and tolerating water stress by continuing growth at dehydration or surviving severe dehydration.

The physiology of the crop plant responses to drought stress has been classified into two domains: (a) appositive carbon balance is maintained by the plant under moderate stress so that resistant genotypes achieve a greater net gain of carbon than susceptible ones and have relatively better yield while (b) a net loss of carbon takes place under severe stress and growth stops and plants are hardly survive.

6.4 Photosynthesis under Drought Stress

Plant growth and development are dependent on environmental factors which determine the types of vegetative cover from arid to rainforest. Low rainfall along with low moisture retaining capacity of the soils causes drought stress in plants. Under drought conditions, root growth is limited to a greater extent as compared to shoot growth. The moisture stress conditions causes decreases in growth rate of assimilatory surface followed by photosynthesis finally results in reduction of productivity due to reduction in CO_2 assimilation.

Closing of stomata reduces net CO_2 assimilation rate (A) of leaf during water stress which can be expressed as:

$$A = g_c (C_a - C_i) \qquad\qquad ....(1)$$

Where, g_c is the leaf conductance for CO_2

C_a and C_i are concentrations in the ambient air and at the mesophyll cell wall

Decline in rate of A when gc is decreasing due to direct effect of the constraints on photosynthesis mechanisms. Decreases in (Ca – Ci) affect internal CO_2 concentration (Ci).

At water stress Ci either remained constant or decreased and chloroplast is isolated from wilted leaves and thus impairs PS II activity due to inhibition in the thylakoid ATP-syntheses and thus photosynthetic apparatus is ceased.

The above hypothesis is not accepted widely due to value of Ci is misleading at the time of patchy stomatal closure during moisture stress as well as lack in considering cuticular transpiration and also due to the fact that even Ci rises when CO_2 uptake and rate of transpiration are low in plants.

Besides moisture stress, plants should also have quality to cope with high light and mechanism to inhibit excess excitation energy from arriving at the reaction centres of photosystems since drought is associated with high solar radiation and high temperatures. Changes in environmental parameters directly affect stomatal closure, osmotic adjustment and decreases in shoot growth compared to roots, tends to maintain water content in cell of the leaves where the photosynthetic apparatus usually damages at 0 to 30 per cent water deficit.

6.5 Drought Resistancy in Photosynthetic Mechanisms

The resistance to water loss and capacity for osmotic adjustment differ in hydrophytes, mesophytes and xerophytes but the sensitivity of photosynthetic mechanisms to water loss does not differ in all these species. A marked reduction in photosynthetic biochemistry has also been recorded at 30 per cent water deficit in C3 plants in which role of stomata is more vital. Some of the studies recorded that under water deficit conditions:

☆ Ratio of PGA to triose- phosphate is decreased

☆ Ratio of triose-phosphate to ribulose 1, 5 biphosphate (RuBP) is decreased

☆ Photosynthetic carbon reduction (PCR) cycle and RuBP regeneration are not affected

☆ Rubisco is also not affected by water deficit.

6.6 CO_2 Concentration inside Chloroplast at Drought

Photosynthetic apparatus is resistant to desiccation and hence, CO_2 concentration inside the chloroplast during mild drought may be low which results an increase in Rubisco activity. Thus decline in rate of photosynthesis during leaf dehydration is only due to decline in CO_2 concentration inside the chloroplast. The following results may occur at drought.

  i. Drought causes an increase in the partitioning of fixed photosynthates to sucrose and starch synthesis at low CO_2 is consistent.

    ii. Decreases in sucrose-phosphate-synthesis (SPS) activity during moisture stress which further justify low CO_2 concentration within the chloroplast.

    iii. Decline in nitrate-reductase (NR) activity may also occur during moisture stress which usually increases NO_3^- levels in the herbaceous plants and closes the stomata to prevent transpiration.

6.6 Maintaining Plant-Water Content during Soil Drying

Reduction in rate of respiration, damages in photosynthetic apparatus, change in leaf movement and thus reduction in CO_2 fixation are quite apparent under water stress conditions.

Reduction in Transpiration

At drying soils, a root tip signal is transmitted to the leaf resulting in closure of stomata, because ABA inversely increases with,

Leaf Temperature

pH of the xylem sap

Leaf water potential

Degradation of Translocated ABA within Mesophyll Cells

Stomatal conductance also responds quickly and reversibly to changes in vapour pressure difference between leaf and air which varies from species to species and water status within the species. An increase in ABA concentration in tissue can induce an osmotic adjustment, thus decreases the osmotic potential in the leaf which results in closing of stomata.

6.7 Light Utilization under Drought

Photosynthetic apparatus is damaged when the leaves are exposed to high light during water deficit High light intensity at drought is more damaging that affect adversely to PS II photochemistry. The following strategies can represent an overall reduction of the primary quinine electron acetones of PSII QA (Quinine acetones).

6.7.1 Mechanisms which Prevent the Absorbed Light from being Used for Photochemistry

Mechanisms which are able to consume the reducing power generated by PS II [Though, both PS I and PS II form NADPH but in PS I, NADP is reduced to NADPH directly, where as in PS II H⁺ is taken from H_2O to form NADPH after passing through plastoquinone (PQ), a primary quinine electron acceptor. Thus, electron and proton transports which are a pH dependent process combine with ADP + Pi to form ATP and thus ATP and NADPH in combination with CO_2 and O_2 involve in Calvine Cycle to synthesize glycolate, triose- phosphate and starch].

Thus leaf movement and curling and deposition of waxes on the leaf hair may be an outcome of waxes on the leaf hair is the result of first mechanism while thermal dissipation at the PSII level belongs to second mechanism while photosynthesis and photorespiration belongs to third mechanism.

6.8 Leaf Movement and Orientation

A rapid and reversible leaf movement is a cause of direct solar radiation which controls the loss of leaf energy budget is termed as heliotropism Orientation in leaves affects photosynthetic activity, transpiration rate and variation in temperature.

Heliotropism is generally found in legumes but it may also occur in Malvaceae, Abiaceae and Oxalidace families. The motor organ for leaf movement is the pulvinus or can happen even in its absence. Paraheliotropic leaf movement in siratro (*Macroptilium atropurpureum*) reduces incidences of light levels. Such movement is also termed as xeric movement which occurs in *Leucaena* during scorching sunlight and saturation of photosynthesis.

Maximum quantum yield of O_2 evolution is measured in planofix leaves since there is an absence of heliotropic movements while erectophyll leaves of the desert shrubs (*Hymenoclea salsola* and *Senecio douglasic*) due to their xeric movement remained constant at 0.5 to -3MPa water potentials. It appears that "the near vertical orientation of leaves serves in maintaining the photosynthetic structures during drought spam.

6.9 Light and CO_2 Utilization during Drought

The gross CO_2 uptake decreases to about 40 per cent in bean and 60 per cent in maize leaves during dehydration. Since fraction of absorbed light energy which is not used in photochemistry increases from 55 to 90 per cent which results decreases in 26 -300 ppm CO_2 m^{-2}S^{-1} during dehydration in both legume (French bean) and cereal (maize).

Contribution of photosynthetic O_2 and CO_2 fixation to energy dissipation decreases from about 40 per cent before dehydration to about 20 per cent after dehydration. Even photorespiration does not protect the photosynthetic apparatus against high light damage.

Finally, it is confirmed that at desiccation, soluble sugars interact with the polar head groups and replace the water molecules. Phospholipid molecules largely retain the original spacing between one other. At moisture level of lower than 0.3 g H_2O g^{-1} dry weight the water dissipates from the water shell of macromolecules and thus the hydrophobic effect responsible for structure and function is lost.

6.10 Water Stress and Nutrient Uptake

Water and nutrient availability is one of the major phenomenons after other physiological functions. Many nutrient elements are actively taken by the plants, decreases under water stress conditions. The roles of different essential plant nutrient elements are well established in regulation of plant metabolism. Majority of studies reported that mineral uptake can decrease when water stress intensity is increased.

In plant like soybean, N uptake is decreased under water stress conditions. In cotton, higher water stress decreases nutrient acquisition due to decrease in transpiration rates resulting loss in active transport and membrane permeability. In some other studies, increase in uptake of nitrogen, K$^+$. Ca^{2+}, Mg^{2+}, Na$^+$ and Cl$^-$ while reduction in uptake of phosphorus an d iron were recorded. Several reports stated that water stress mostly results reduction in recovery of nutrients as such phosphorus, K$^+$, Mg^{2+} and Ca^{2+} in some crops while Ca^{2+}, Fe^{2+} Mg^{2+}, nitrogen, phosphorus and K$^+$ in some other species like *Spartina alterniflora*, Fe^{2+}, Zn^{2+} and Cu^{2+} in sweet corn, Fe^{2+}, K$^+$ and Cu^{2+} in *Dalbergia sissoo*.

Nutrient uptake under different moisture levels also varies as per plant species as well as due to varietal differences. In one of the study, availability of nutrient elements decreases in order of Fe^{2+}> K^{2+}> Mn^{2+}> Zn^{2+}>Mg^{2+}> Ca^{2+} from 25 to 6 per cent. Interactions among nutrients are also responsible for uptake under moisture stress conditions. In herbage species, the uptake and solubility of nutrient elements depressed but Ca/K and Ca/P ratios increased under water stressed conditions. From the different reports, the impairment in active transport systems in drought conditions as well as due to variations in nature of different nutrient elements is responsible for inhibition in availability and uptake of nutrient elements. Thus it is very clear that a wide variation in morphological, physiological and biochemical indices are responsible for drought resistance in crops and varieties.

6.11 Drought Stress and Water-Use-Efficiency

The selection of drought resistant crops and their varieties should be done under specific field conditions keeping the following line of action.

  i. Realistic soil conditions

  ii. Testing under both adequate and inadequate water conditions

  iii. Assessing crop failure in the area with reasons and

  iv. Selecting a limited number of traits for improvement.

Proper knowledge of soil conditions, climatic variations, physiological and biological characteristics of the crop species is needed for meeting the above requirement and success there in.

Since water is the most important input for the dryland area hence, testing the water use efficiency (WUE) of the crop can be an ideal parameter to assess the resistancy under limited water conditions.

$$WUD = \frac{D}{W} \qquad\qquad ....(2)$$

where,

D = The mass of dry matter produced (usually above ground)/Yield

W = Mass/volume of water used (evapotranspiration).

The WUE does not vary on the changes in availability of water but differs as per species. Among the three types of photosynthetically classified species (C3, C4 and CAM), CAM species Cactus has the highest WUE (20 g above ground dry matter/kg water followed by C4 species (sorghum, sugarcane, maize and several semi-arid perennial grass species (3-5 g dry matter/kg water) and least of C3 species (Rice, wheat and annual grain legumes (2-3 g dry matter/kg water).

Since loss of water is the sum of evaporation from the open land surface and from transpiration by the plant hence, the WUE may be modified as,

$$WUD = \frac{D}{D\,max} = \frac{W}{W\,max}$$

Where, D/D max is the fractional dry mass expressed in relative to the maximum dry mass produced with optimum water D max,

W/W max is the fractional water use expressed relative to the maximum evaporation W max that could occur with optimum water.

Among these two hypothesis referred at No.1 is appeared to be more justified since farm income for an irrigated crop is based on the absolute economic yield rather than normalized yield and the response on absolute quantity of water supplied.

Chapter 7

Agronomy of Drought Stress

Present day crop management system in dryland or water stress conditions either in low rainfall area or high rainfall area with low soil water retaining capacity is based on low inputs combines with soil and water conservation practices with risk minimizing strategies. Water scarcity is the prime limiting factor. Besides this dryland system also takes an account of poor soil fertility status, weeds and other biotic stresses which reduce the water use efficiency.

Watershed management under dryland environment is the life line for the success of a sustainable farming system since seasonally or occasional dry regions are the major constraint to agricultural development. Therefore, watershed should be selected as a unit for resource development in any given region. Watershed farming system involves optimum utilization of the catchment precipitation through an efficient use of water and soil for crop, dairy and other enterprises. This can be implemented as:

    i. Directly through infiltration of rain water

   ii. Runoff harvesting, storage and recycling and

  iii. Deep percolation and recovery from wells, tanks and lakes.

The land utilization should be based on their capability classes. Generally such land has been classified into eight types (from Class I to VIII) to be engaged under different systems (Table 7.1).

Table 7.1: Land Capability Classes and their Suggested Use.

Land Capability Classes	System
I	Intensive cropping
II	Crops and Temporary Pasture(Leys)
Iii	Temporary crops and permanent pasture
IV	Very limited crop and permanent pasture
V	Fruit trees and woodland
VI	Moderate grazing and forestry
VII	Permanent forest and restricted grazing
VIII	Wild life and recreation

The sustainability of a dryland farming system may involve the following principles.

i. Improved soil and water conservation practices with minimum tillage in a watershed area.

ii. Optimization of fit between crop growth cycle and the available moisture

iii. Weed control.

iv. Soil fertility management

v. Control of soil biotic stress which is responsible for restricting root development

vi. Optimization of plant geometry as per soil moisture regimes

vii. Crop diversification rather than mono cropping

viii. Integration of improved forage-livestock-grain crop system

ix.Feasibility of rain by cloud seeding

x. Quick in operation just after rain say, seeding, top dressing and weeding etc.

7.1 Soil and Water Conservation

Increasing storage of soil moisture by the rotational fallow system with or without conservation tillage is an essential practice in dryland farming elsewhere. The benefit of alternate fallow and conservation tillage to increase soil moisture for the crops depends on:

☆ Soil water-retaining capacity

☆ Climate

☆ Topography and

☆ Management practices

Conservation tillage involves minimum tillage operations to conserve soil structure and to maintain ground cover by mulch to reduce water runoff and evaporation and increase infiltration. Other practices like deep tillage to break hard pan to reduce bulk density and soil pitting or formation of small depression at close proximity to store water and reduce runoff from rain storm are also followed.

7.2 Water Harvesting

The basic water harvesting systems involve an external contributing area to reduce runoff. This area is treated physically and chemically for the maximizing runoff. The water is diverted into a receiving area comprising of cultivated plots, individual trees or small terraces. The water contributing area includes crop land (a system sometimes referred to as "conservation bench terrace") or outside the field in the natural watershed system.

7.3 Diversification of Farming

It is an effective approach to reduce the risk associated with farming in unpredictable climate. Reduced diversification to even mono-cropping is possible only with a high level of control over the crop climatic conditions. Diversification of cropping to reduce risk particularly stands true in dryland situations. It is achieved on several levels for the case of traditional rain-fed rice in Eastern India.

Spatial Classification of Fields

The land is divided into plots as per topography and soil and hydraulic properties. Some plots may be prone to water logging and other without water.

Crop Diversification

Mixed or inter cropping is done with different proportion of rows of the associated component crops to cover risk.

Temporal diversification under which the component crops are sown in rows at different date intervals. These crops are sown before or just at the onset of pre-monsoon rain. It covers the risk of unpredictable rain and gives yields of at least from one or two dates of seeded crops.

7.4 Components of Drought Resistance

Dehydration avoidance and **dehydration tolerance** are the two major path ways that govern drought resistance in crop plants. Dehydration avoidance is the capacity to avoid plant tissues and cells dehydration under drought stress while dehydration tolerance is the capacity to sustain when the plant is dehydrated. Plant survival can be conditioned by either of the two. During the impact of stress, moisture stress signals the expression certain stress responsive genes, which are responsible for a chain of events and gene "networking" may be identified stress and soil conditions can affect deep rooted genotypes that may have an advantage over shallow rooted genotypes to extract moisture present in the lower strata.

7.7 Dehydration Avoidance

Leaf area index (LAI) has a marked control over water-use by small plants with reduced leaf area by performing xeric movement helps the plant to withstand water stress. Cultivars developed to grow under dry stress conditions have potential for higher water-use-efficiency with moderate plant size results in higher yield performance. It has also observed that early seeding with better canopy cover restrict the evaporation. Early onset of reproductive phase helps in escaping drought.

7.8 Root System

Root length and density are the main traits which meet the transpiration demand. Such characteristic in upland rice and grasses with less tillers production helps in deeper soil penetration. The crops with capacity to break the hard pan of soil are better equipped in extracting moisture from lower layers since it also facilitates the water entry. Hence, short growth period, smaller leaf area, higher hydraulic resistance of fine roots with smaller diameter are favourable to withstand during drought stress.

7.9 Plant Canopy/Surface

Plant surface structure determines the reflective properties of leaves and their resistance to transpiration. It absorbs light energy for CO_2 fixation and reflects excess radiation. Leaf resistance to transpiration is dependent on the size, shape, number and forms of stomata including leaf hairs. Second to stomata, it is the cuticle, which transpire the water. Hydraulic conductivity of the cuticle is determined by presence of wax and gummy materials embedded in the cuticle. High cuticular permeability affects transpiration and also directly affects loss of water from guard cells. Besides these, presence of pubescence on leaves increases reflectance from the leaves surface is one of the characteristics of desert xerophytic species. It increases reflectance from the leaf within the light range of 400-700 nm and sometimes up to 900 nm, results in lower leaf temperatures under scorching sun light. Colour of the leaf affects thermal properties of the leaf. As such, the yellow colour cultivars tend to perform better under drought stress as compared to green colour cultivars.

7.10 Osmotic Adjustment

It facilitates in maintaining cellular turgor pressure at a given leaf water potential and thus delay wilting. At drought stress different solutes (potassium, sugars, poly-sugars, amino acids and glycinebetaine) of the plant accumulate in cells and thus the osmotic potential of the tissue is reduced. Normally the rate of plant dehydration should not be faster than about 0.1 MPa/day. It varies as per species and cultivars. It should be about 0.3 MPa but in cereals it is from 1.5- 2.0MPa. This is the reason that osmotic solutes are also known as "protectants" to drought stress. Therefore, extensive bio-engineering efforts are being made to evolve transgenic model plants that can accumulate osmolytes.

7.11 Dehydration Tolerance

Cellular and molecular adaptive responses one or more of the following functions.

☆ Reduce whole plant growth in order to reduce plant water-use

☆ Reduce the rate of cellular water loss and retain cellular hydration and

☆ Protect various cellular structures and functions as cells desiccate.

Though, the modern genomic tools are trying hundreds of genes which are up or down regulated in response to plant tissue water loss but final result is yet to come.

7.12 Stem Reserve Utilization

The first source of carbon for grain filling is fixation by the intercepting viable green leaf area but this source diminishes due to stress conditions. Small grains and cereal stems and several other crops store carbohydrates in the form of glucose, sucrose, fructose and fructans or starch. These are stored as total non-structural carbohydrates or water soluble carbohydrates and are available for translocation to the grain.

Usage of stem reserve depends on the available storage and the rate and duration of mobilization of storage to grain. The storage and transfer to grains also a subject of genotypes, growing conditions as well as on length and density of the stem or say, sink size. In wheat, dwarfing genes Rht1 and Rht2 have limitation to transfer the materials from stem to grain as it reserve about one third of materials due to shortening of stem. This might be the reason of its susceptibility to drought stress.

Under heat stress starch synthesis in wheat grain may be restricted by a thermo labile enzyme and available stem reserves may not be in demand by grain. Therefore, heat tolerant starch synthesis is essential for grain filling under heat stress. Role of hormones also can not be ruled out. Stem reserve transfer is a main source of carbon for grain filling under any stress including heat stress.

7.13 Stability of Cellular Membrane

Cellular membrane is the central site for function associated with membrane bound enzymes and transport of water and solutes. Cellular membrane function under desiccation and heat stress is the plant previous exposure to moderate stress signal of acclimation effect that is expressed in increased membrane stability under stress is the most significant factor. Membrane stability is assessed by measuring the leakage of cellular electrolytes under stress.

Water passage is regulated by both plasma membrane and tonoplast is vital to cell life and specific proteins. These "water –channel" proteins, are also termed as **acquaporins,** which response to signals and "molecular switches" These pores are very selective to water and play an important role in cellular water relation in response to plant water deficit underwater stress.

7.14 Antioxidation

Oxidative stress is used to describe a state of damage caused by reactive oxygen species. These are the active material naturally occurring in all organisms which detoxify free radicals. Superoxide dismutase (SOD), catalase, glutathione reductase or ascorbate peroxidase are known as antioxidant particularly SOD converts the O_2 to H_2O_2 and catalyze converts H_2O_2 to molecular oxygen.

Drought as well as other stresses cause oxidative stress in plants and antioxidant activity is important for the protection of metabolism under stress. Drought induces oxidative stress related to genes and is associated with increased levels of various antioxidants in plants needs development of transgenic plants.

7.15 Abscisic Acid (ABA) Accumulation

Accumulation of ABA in plant affects stomatal closure and reduced transpiration As such transgenic model species like tobacco is engineered to over-produce ABA to maintain turgor during water stress. However, ABA has a number of negative effects on plants (Table 7.2).

Table 7.2: Effects of ABA on Plant Process Involved with Growth and Reproduction.

Growth	Effect	Growth	Effect
General growth	Inhibition	Cell division	Decrease
Cell expansion	Decrease	Leaf initiation	Inhibition
Germination	Decrease	Root growth	Decrease
Tillering	Decrease	Dormancy	Improved
Reproduction	Effect	Reproduction	Effect
Flowering (annuals)	Advance	Flower induction	Inhibition
Flower abscission	Increase	Pollen viability	Decrease
Seed setting	Decrease		

7.16 Stress Proteins

Stress proteins are a large group which are induced by variations in environmental and biotic stress in different organisms ranging from prokaryotes to man. A group of small molecular weight protein is developmentally regulated in growing seed. Their accumulation at embryo development has a role in protecting the embryo as the grain matures and desiccates during maturation, even at 10 per cent water content, is defined as "late embryogenesis abundant" (LEA).

LEA family of stress protein may have a role in drought and osmotic stress resistance and may also act as molecular chaperson and in that respect they are very similar to low molecular weight heat shock protein (HSP) which may conserve proteins during stress as work with transgenic gene indicated.

7.17 Drought Stress and Plant Phenology

Generally, early flowering is defined as short growth duration genotype has an important attribute of "drought escape." Longer growth duration in both determinate and indeterminate plants would improve the probability for regrowth upon recovery under reoccurring of favourable conditions. The short duration cultivar delay (rice) their flowering when stress occurs before flowering. The rate of delay is a function of plant water deficit and probably ABA signaling. The rate of change in flowering time under stress can be taken as an index of genotypic rate of stress in the field.

7.18 Water-Use-Efficiency (WUE)

This term is used for higher production in a given amount of water as "more crop per drop". Sometimes high WUE results in "less drop per crop". WUE was developed by agricultural engineers as a ratio between yield and irrigation water to asses the cost. Later on it is used by the agronomist usually in dryland-rainfed crop production. Crop physiologist found the term useful at the leaf level in studies of gas exchange where WUE (*i.e.*"transpiration ratio") is defined as the ratio of carbon fixation to transpiration.

The breakthrough came with the development of better understanding of stomatal dynamics; gas exchange and photosystem function, leading to the carbon isotope discrimination (delta) assay as a heritable marker for WUE at the whole plant level. In majority of cases low carbon isotope discrimination (low delta) as measured in the grain or leaves is found to be well correlated with high WUE across variable genetic materials and vice versa. Depending on the crop growing conditions, the relation between WUE and yield was sometimes positive and sometimes negative. Plant breeders discussing carbon isotope discrimination and WUE expressed confusion on two accounts;

Under what environmental conditions selection for carbon isotope discrimination is expected to result in yield gain.

Which direction should selection be made, high (low delta) or low (high delta) WUE.

Higher WUE is derived from a reduction in water use rather than from an increase in production. Reduced water-use under dryland conditions is contradictory to productivity since genotypes of high WUE under drought stress tend to be less productive under stress with few exceptions hence, it appeared that *the target of plant breeding for water limited environments is effective use of water (EUW) rather than WUE.*

7.19 Photosynthesis Under Drought Stress

Though, this is discussed in details in climatic stress however, performance of different photosynthetic metabolisms systems under drought stress conditions are being discussed here. The C4 system has a superior productivity at a given water-use. Such as in maize, sorghum, bamboos, pearl millet and a number of forage grasses have a pumping mechanism or "Kranz leaf anatomy" that re-capture the CO_2.and hence CO_2 concentration in bundle-sheath is higher, results in higher assimilation with high WUE. Decline in C4 photosynthesis with water deficit is due to metabolic limitations to CO_2 fixation, whereas, in C3 species with better stomatal conductance are more efficient in recovering under re-watering conditions. The high WUE C4 species are metabolically more sensitive to drought than the C3 species and recover more slowly from drought.

Attempt is being made in converting rice plant from C3 biochemistry to C4 to give higher leaf productivity to per unit transpiration since it has the lowest WUE.

7.20 Selection for Drought Stress

The comparative performance of different genotypes under variable environmental conditions is generally known as genotypic x environment (GxE) interaction or genotypic x phenotypic interaction for their yield performance. It is referred to as stability parameters in which usually more than 10 genotypes performance in their yields during minimum of 3 years or at 3 locations having variable environment are tested. The genotypes are screened on the basis of criterions proposed by Eberhert and Russell (1969) which states that;

☆ The yield of the genotype must be higher than the grand mean yield

☆ The regression coefficient (b) should be to unity or nearer to unity (1).

☆ The deviation from regression line should be zero or nearer to zero.

A genotype giving highest yield but having high value for deviation from the regression line is said to be ideal under favourable conditions while a genotype that gives yield just above to grand mean with regression coefficient one or nearer to one with almost zero value for deviation is the best suited under variable environmental conditions. The other that produced yield just equal or just below the grand mean with regression coefficient to unity and zero or almost zero value for deviation from regression line is definitely suited to grow under stress conditions.

Though, sometimes the stress resistant index (RI) proposed by Fischer and Maurer can also be computed however, under this conditions the genotypes are tested under two different conditions of stress and non-stress environments separately and following statistical model is followed.

$$RI = (Gs/Gn)/(Ms/Mn)$$

where, Gs is the genotype yield under stress

Gn is the genotype yield under non-stress

Ms is grand mean yield of all genotypes under stress and

Mn is the grand mean yield of all genotypes under non-stress.

Values above 1 indicate a relative resistance as compared with the mean of the genotypes, this index is flawed due to the fact that it is influenced by yield potential. It is also true that high yielding genotypes require high management conditions while genotypes grown under stress or limited input conditions are less productive but can sustain in a better way.

Chapter 8
Nutrient Stress

Though, increases in temperature accelerate the decomposition of soils organic matter and release of nutrient however, fast loss in soil fertility due to high rate of mineralization is equally serious along with reduction in the population of soils micro-organisms. The countries of whose soils are poor in organic matter may suffer more due to burning effects of organic carbon and release of CO_2 under high UV-B radiation and temperature. Since soil organic matter is the storehouse of all the nutrient elements hence status of organic matter in soil is an indicator to quantify the soil fertility which is going to be deteriorated due to continuous cropping without incorporation of organic materials in the soil due to loss in production of farm yard manures as the drastic reduction in livestock population on one hand and burning of organic carbon due to global warming.

8.1 Global warming and Soil Organic Matter

The carbon cycle is consisted of 5 major or global carbon pools are interconnected and carbon circulates among them (Figure 8.1)

The increase in atmospheric C pool is responsible for increases in atmospheric CO_2 and CH_4 gases concentration and C cycle has direct effect on water cycle and increases in other climatic parameters.

Biotic pool	Geological pool
560 Pg (+2 Pg/yr)	5,000 Pg
	Atmospheric pool
	760 Pg (+3.3 Pg/yr)
Pedologic pool	Oceanic pool
2,500 Pg	38,000 Pg (+2.2 Pg/yr)

Figure 8.1: Human-Induced Changes in Global Carbon (C) Pools and Fluxes.

The carbon pool in soils of the tropics is composed of 496 Pg of soil organic carbon (SOC) and 210 Pg of soil inorganic carbon (SIC). Historic land-use soil/crop management practices have resulted depletion in carbon pool by17-39 Pg. Strong and extreme forms of soil degradation by water erosion is 150 Mha,15 Mha by wind erosion, 115 Mha due to loss of nutrients, 43 Mha through salinization and 33 Mha as physical degradation.

Judicious cropping with high yield harvest has also caused drastic loss in original SOC pool in the tropics. Soil carbon sequestration will result in improvement in soil fertility and thus increase in productivity. Restoration of degraded soils with addition of SOC may also increase soil inorganic carbon in irrigated soils The large potential of soil carbon sequestration in the tropics over a 50-year period is composed of three main components.

(i) Restoration of degraded soils and ecosystems including

    (a) Erosion control with potential of 10- 25 PG

    (b) Restoration of strong and extremely degraded soils at 5.7 - 10.8 Pg, and

    (c) Bio-fuel production on degraded soils at 57.5 – 115.0 Pg

(ii) Adoption of recommended agricultural practices on

    (a) Crop land at 2.2 – 4.1 Pg, and

    (b) On pasture land at 6.0 – 12.0 Pg

(iii) Soil inorganic carbon sequestration through leaching of bicarbonates in irrigated soils at 2.2 – 8.7 Pg.

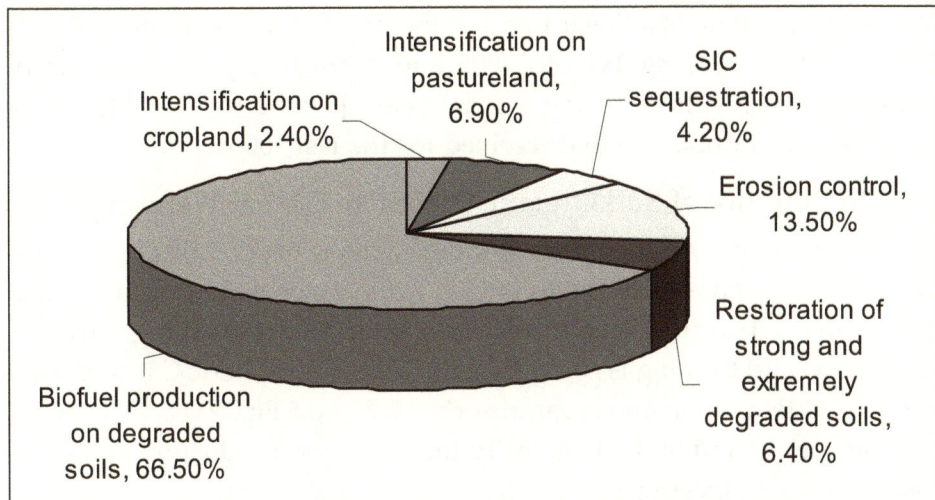

Figure 8.2: Relative Contribution of Different Options for C Sequestration and Emission Reduction with a Total Potential of 81-187 Pg over a 50 Years.

Total potential of C sequestration and bio-fuel offset over a 50 year period (Figure 8.2) is 83.6-175.7 pg (mean of 129.7Pg) at the rate of 1.7-3.5 Pg C per year (mean of 2.6 Pg per year). Realization of this vast potential is a challenge for researchers, extension agents, land managers and policymakers. The relative contribution of different components shown in indicate that 66.5 per cent of the 129.7 Pg of sequestration potential lies in bio-fuel production and fossil fuel offset, 13.5 per cent in erosion control, 6.9 per cent in intensification of pastureland, 6.4 per cent in restoration of degraded soils, 4.2 per cent in SIC sequestration, and 2.4 per cent on intensification of cropland, Although the potential of soil C sequestration in the tropics is large, the realization of this potential poses a major challenge. For instance, major difficulties in meeting this challenge include lack of research data, weak institutions, and poor logistic support. There is a strong need to develop a coordinated program in soil C sequestration as this is truly a win-win situation.

Organic carbon content in soil is double to its concentration in atmosphere and almost thrice to what vegetation contains. Therefore, any loss in soil organic matter (SOM) has significant effect on rising in CO_2 concentration in atmosphere and simultaneous enhancement in global temperature. There is a linear trend in increases in temperature of $0.74^{0}C$ $(0.56 - 0.92^{0}C)$ during 1906 to 2005 and accordingly a linear rate of warming has resulted during last 50 years $[0.13^{0}C$ $(0.10 - 0.16^{0}C)/10$ years is almost double to that received for the last 100 years.

8.1.1 Importance of Soil Organic Carbon to Global Warming

The organic carbon content in global soils is about 1,500 Pg (1 Pg = 1Gt = 10^{15} g) and annual fluxes of CO_2 from atmosphere to land. Global net primary productivity (NPP) and land to atmosphere through respiration and burning is 60 Pg C/year. Thus atmospheric C is increased @ 3.2 +- 0.1 Pg C/year and ocean absorbed 2.3 +- 0.8 Pg C/year. Therefore, soil has lost 40 and 90 Pg C globally through crop production and other disturbances, respectively.

Even a small change in soil organic carbon (SOC) brings a very high increase in atmospheric CO_2 concentration and subsequently rise in temperature brings about a drastic change in ecosystem.

8.1.2 Global Warming and Decomposition of SOM

Photosynthesis and respiration guide the balance of carbon between soils and atmosphere is regulated by temperature. Some of the factors proposed to be responsible for organic carbon stabilization are:

(i) Microaggregation (53-250 mm) formation within macroaggregates

(ii) Physical binding with clay and silt particles and

(iii) Bio-chemical formation of recalcitrant soil organic compounds

Loss of SOM is less in alkaline soils due to slow decomposition rate in presence of hydroxides, oxides, phosphates and inorganic carbon, Enhancement in the formation of soil aggregates is also influenced by soil organic carbon, calcium and magnesium ions.

In wetland, low temperature, anaerobic conditions, poor substrate quality, low pH, and low nutrient availability stabilize the soil organic matter.

Loss in SOM may bring changes in water balance, atmospheric composition of $O_2 : CO_2$ concentration and land use system. Hence, increase in biomass production, slowing in the rate of OM decomposition and check in industrialization may keep a balance between SOM and atmospheric CO_2 concentration.

8.2 Effect of Elevated Temperature

Soil temperature is directly influenced by air temperature which affects root growth and development. Low soil temperature reduces the solubility of several nutrients to cause deficiency. As a result reduction in dry matter production, turning of young leaves to dark green and anthocyanin production may occur.

The carbon sink capacity of soils degraded from 50-66 per cent leading to loss of 42-78 Giga tons of carbon. An increase in 1°C in temperature results 10 per cent loss of soil organic carbon leading to enhance in global warming to a greater extent while an increase of 1 ton of soil carbon-pool of a degraded soil may increase crop yield by 20-40 kg/ha of wheat, 10-20 kg/ha maize and 0.5-1 kg/ha cowpea. The rate of decomposition of organic carbon increases with temperature at 0°C with Q10 of almost 8.

Increases in soil temperature activate the solubility and availability of nutrient elements. Therefore, at every 10°C (Q10) rise in soil temperature, the nutrient uptake is almost multiplies by 2. Increases in soil temperature are more favourable for winter crops; sesame, oilseed rape, wheat and barley while reduction in nutrient uptake may be noticed in tropical crops; sorghum and some cultivars of tomato. Hence, it appeared that nutrient removal by crops may depend on soil temperature and types of species to be grown. Crops already growing in soil optimal temperature may suffer due to increases in soil temperature. Since roots have lower temperature optima than shoots hence elevation in temperature may be injurious for crops.

Nutrient uptake is also controlled by roots respiratory products, roots membrane temperature, root hydraulic conductivity and chemical forms of the nutrient.

The absorption of P and K by maize roots reached to a maximum at 25 and 35°C whereas roots respiration rate reached to a maximum at 40°C. The Q10 values for P uptake were very high compared to Q10 values for respiration. Therefore, it is difficult to conclude that reduced ion uptake with decreasing temperature and increased ion uptake with increasing temperature, is entirely dependent to drop/rise in respiration.

It appeared that rise in temperature may alter the nutrient relation by affecting the fluidity of membranes and their protein transporters. It is expected that short-term changes in soil temperatures will directly affect the uptake by changing membrane properties and in turn the kinetic of enzyme function.

Root hydraulic conductivity affects the rate of water passage through the plant shoot and may also affect the transport of nutrients in soil solution on to the surface of membrane. Therefore, increases in soil temperature increases the root hydraulic conductivity. In barley, root hydraulic conductivity increased at 25°C and in wheat, increase in root temperature from 20 to 30°C increases the root hydraulic conductivity from 200 to 400 per cent. However, in some other studies, rise in soil temperature reduced the hydraulic conductivity due to absence of optimum availability of the nutrient which resulted in appearance of nutrient deficiency symptoms.

Chemical forms of nutrient have also a close relationship with soil temperature. At high soil temperature, NO_3^- nitrogen form is preferred over NH_4 nitrogen and accordingly at low temperature NH_4-N is preferred hence application of NH_4 form of N at high soil temperature causes toxicity. Hence, NH_4 form of N-fertilizers should be applied in winter season and NO_3 form in warm season.

8.3 Effect of Elevated CO_2 Concentration

Higher $[CO_2]$ reduces the respiration per unit of root mass and thus the uptake of nutrients. The mineral concentration in plant decreases because even at optimum supply of nutrient the uptake is reduced. Consisted effect on growth at elevated $[CO_2]$ on ion uptake kinetics does not appear. As such, exposure to high $[CO_2]$ does not change the kinetics of NO_3 or NH_4 uptake in soybean and sorghum.

Changes in transportation of highly mobile N and high $[CO_2]$ might be due to changes in plant water use and thus lowering in concentration of plant tissues. Furthermore, reduction in stomatal conductance and transpiration at high $[CO_2]$ is very common. Reduction in bulk flow of water from soils to root and less in presence of nutrient at the root surface also reduces the nutrient content in plants.

Since, Hill and Rubisco activities are strongly related to leaf-N as well as for photosynthesis I and II (PS I and PS II) and other pigment – protein complexes (Cytochrome) hence photosynthesis is directly affected by N availability. More than 50 per cent of the leaf – N and larger portion of the remainder is directly and indirectly plays role in photosynthesis. Increases in CO_2 concentration though increases the bio-mass but reduces the quality due to reduction in plant-N which is more influenced in C_4 as compared to C_3 species. The photosynthesis rate to per unit of N (photosynthetic N-use efficiency, PNUE) at the growth irradiance is the highest in the leaves with low N concentration due to higher degree of utilization of the photosynthetic apparatus. The CO_2 fixation under N constraint is the down related process due to decline in Rubisco, chlorophyll and stomatal conductance.

N-concentration per unit leaf area of leaves facing the sun-light may be higher to those under the shade down the canopy. Therefore, upper younger leaves with higher N content may have higher CO_2-assimilation. Plants growing with high N application under low light intensity are low in dry matter content due to utilization of only a fraction of the photosynthetic capacity.

Uptake of immobile nutrients (P and Zn) at high $[CO_2]$ may be increased due to increases in root density and solubility of these nutrients at high soil temperature. This might be one of the reason that succeeding rice may response well on the residual-P, applied in preceding wheat. Variations in uptake of different forms of nutrient also differ from species to species. In half a dozen of grasses, NO_3^- uptake was not affected at high $[CO_2]$ but uptake of NH_4^+ was slowed. In some of the crops like soybean and grain sorghum the rates of both NH_4^+ and NH_4^+ uptake were hardly affected by elevation in $[CO_2]$ which indicated the preferences of forms for ions by plant species. However, elevation in $[CO_2]$ stimulates the N-fixation due to higher rate of photosynthesis and supply of respiratory substrate to roots and nodules.

The plant N demand at higher $[CO_2]$ is reduced due to limitation in photosynthetic rates compared to carboxylation by rubisco. Therefore, even at higher supply of N, the demand is restricted resulting in reduced N-concentration in tissues at elevated $[CO_2]$. Some compensatory mechanisms may explore the possibility to enhance nutrients uptake through better root development. Exudation of organic acids, sugars and amino acids by roots may stimulate nutrient availability by accelerating decomposition and mineralization. This is a positive sign that there may not be an adverse effect on plant nutrition and yield under elevated [CO_2] in future, of course the protein concentrations may reduced markedly.

Some of the studies on different crop species have indicated that in soybean causes of failure in solar-tracking in P-deficient soils. Stomatal conductance in rice and maize under both N and P deficiencies has been well established. In white pine (*Pinus strobes*) and *Pinus taeda* L. N and P nutrition is positively correlated with photosynthesis and magnitude of photosynthetic response to high CO_2 concentration is of course dependent on soil fertility. Plants growing under nutrients stress conditions give very less response to elevated CO_2 than those well-nourished. In *Gossypium hirsutum* leaves net CO_2 uptake reduced at low N, P and K concentrations because of thin leaves formation with lower chlorophyll content decrease the fixation of CO_2.

Soil with low available-P, photosynthesis may show little correlation with tissue-N, but a strong correlation with tissue P-concentration. Plants grown on low-P usually show low photosynthetic rate due to feedback inhibition caused by low concentration of P in the cytosol or low concentration of Rubisco and other photosynthetic enzymes.

Some scanty information on the effects of micronutrients on photosynthetic reaction also indicated that Mn-deficiency affect PS II reaction center in low chlorophyll containing leaves.

8.4 Effect of Elevated [O_3] on Nutrient Availability

Since atmospheric O_3 does not penetrate the soil surface hence it cannot directly influence the uptake of nutrient by roots however, shoots exposure [O_3] to high [CO_2] can indirectly affect the mineral removal. The uptake of nutrient can be affected by the following ways:

1. There may be competition between K and O_3 due to their presence in stomata and thus O_3 presence may restrict the stomatal movement which is the role of K.

2. Nutritional factors responsible for repair of cells can directly affect plant sensitivity to O_3 injury.

3. Exposure to high O_3 can result to up-or down-regulation of genes, including the gene that code for plant hormones which can affect a wide range of plant development including roots.

4. It diminishes CO_2 fixation, growth and possibly demand for nutrients.

5. Presence of O_3 may affect the mineral nutrition by controlling leaves density in plant canopy.

Incidence and intensity of O_3 damage increased with increasing soil fertility. In more than a dozen experiments, leaf injury was more severe in tobacco fertilized with optimal N to those raised at low N. Several observations have also endorsed that low N fed crops were more resistant to O_3. Plants grown at optimum N produce more rubisco, is targeted by O_3.

In some of the very limited works indicated increasing P level induces O_3 damages. Though, it was not recorded in radish but greater damages in soybean and other bean were observed at low K which was absent at higher K. Crops fed with higher S were more susceptible to O_3 damages due to competition for same site by sulfhydryl group and O_3.

Though, atmospheric O_3 had very little effects on P, K, Ca, Mg, Mn and Na concentration in tissues of wheat but total P and K concentration in flour were decreased by high O_3 but N concentrations were increased in both straw and flour. The some scanty data indicate that O_3 deficient crops contain more nutrients than O_3 rich climate due to obstructions in nutrient uptake from soils.

8.5 Effect of UV-B Radiation

In a few study, increases in duration of UV-B radiation affected the growth and pigment of Scots pine (*Pinus sylestris*) plant, while in some others thickness of leaves was common. In one of the study on wheat in China, the increases in concentrations of N and K in all parts of the plant were observed. Changes in the concentrations of P, Mg, and Zn were varied in tissue-dependent manner; decrease of P in leaves and stems but increases in spikes and grains were noticed. The weight of N, P, K, Mg, Fe and Zn in various parts of the plant decreased except leaf- N weight was increased. UV- B radiation also decreased the concentration of soluble carbohydrates in leaves and increased that of holocellulose and soluble proteins. After 60-100 days decomposition of leaves and stems in the field, enhanced UV-B radiation stimulated the loss of organic carbon and thus loss in nutrient content of soils.

Some of the works done in India on the effect of UV-B radiation on wheat, recommended 50 per cent increase in the level of N, P and K nutrient to avoid the adverse effect of radiation (280-320 nm). At shorter light wavelength, the substantial increase in population density of nematodes at high summer temperature was also recorded.

In the experiment conducted in the greenhouse, the different doses of UV-B radiation applied to the two species *Avena fatua* and *Setaria viridis* induced changes in leaf and plant morphology. It was a decrease of plant height, fresh mass of leaves, shoots and roots as well as leaf area. Besides, it caused the leaf curling in both of the species. The significant differences between *Avena fatua* and *Setaria viridis* in the studied traits were mainly due to the tillering ability of the species. The content of chlorophyll varied considerably. The average values of leaf greenness (SPAD units) for oats were about 43 while for green foxtail 32, respectively. U-VB did not reduce leaf weight ratio, shoot dry matter, shoot to root ratio and leaf area ratio.

Chapter 9
Water Requirement

There are 1,400 mkm³ of water on the earth to cover the earth surface to a depth of some 3 km. 97.25 per cent of water is found in oceans, 2 per cent as glaciers and ice caps. Only 0.7 per cent water is as ground water. Of the 0.05 per cent water 60 per cent is in Lakes, 33 per cent as soil moisture, 6 per cent in atmosphere and 1 per cent in rivers. The average resting time for water in atmosphere is about 10 days and in longest rivers is 20 days or less and for soil moisture is about 30 days.

Table 9.1: Availability and Withdrawal of Water in Continents (m³/person/year).

Continents	Water Availability/capita	Water Withdrawal/capita
Oceania	76,000	907
South America	35,000	476
North and Central America	16,000	1,692
Africa	6,500	244
Europe	4,700	426
Asia	3,400	526

Among the different continent, the availability and consumption of water to per capita (Table 9.1) in African and Asian countries is much below the average. That is why, availability of even fresh water for human consumption in several African and Asian countries (particularly Gulf Nations) is a big problem.

About 500 000 km³ of waterfalls as rain or snowfall of which 113 000 km³ of water are evaporated from the earth's surfaces and vegetation annually. Surface runoff and ground water seepage enter streams and rivers that, in turn, flow downstream in to the oceans. Only 40,000 km³ of water returned to oceans which are transferred annually in form of clouds.

The Earth with presence of water is known as living planet. As such we are searching for water on other planets since life cannot exist without water. Between soil and plant, it is the water that facilitates soil to supply nutrients and water to plants to manufacture food for all livings. Global warming has virtually changed the rainfall pattern and total precipitation round the year. The areas which were traditionally fall under high rainfall zones are facing drought while the semi-arid zones are receiving excess rain or even unpredicted floods in recent years.

The extreme low temperature (- 45°C in Dras of Ladakh, east of J&K) and extreme high temperature (50.6°C in Alwar, Rajasthan) on one hand and there is the highest rainfall of 11,871mm in Mawsynram village of Meghalaya, India recorded the highest rainfall in Asia on the other hand. In a single day, the highest rainfall of 650 mm occurred on 26th July' 2005 in Mumbai killed 900 people. The highest snowfall (8.4m) in Gulmarg in 1967 and 12m snow drifts in February' 2005 killed 200 people.

The extreme drought conditions also brought four times Bengal famine during 1770, 1876-77, 1899 and 1943 in which altogether more than 20m people died. An *El Nino* – Southern Oscillation (ENSO) – related oceanic low pressure convergence center forms; it then continually pulls dry air from Central – Asia desiccating India, the reversed air flow causes India's droughts. Accordingly the sea surface temperature of Indian Ocean in 1997 – 98 raised to 3°C resulted in high evaporation-wet weather in India

[*El Nino southern oscillation: is a global coupled ocean-atmosphere phenomenon. Pacific Ocean signatures, El Nino and La Nino are the major temperature fluctuations in surface water of the tropical eastern pacific ocean. Since the phenomenon occurs around the Christmas time in the pacific ocean off the west coast of South America and therefore, these names which are derived from the Spanish literature for 'the little boy" (El Nino) and "the little girl" (La Nino) are given, refers to the Christ children].*

Changes in climatic indices resulted in severe drought and flood in India of which five times were assessed as extreme (Table 9.2). Besides these, severe drought during 2009 and continuously in passing year 2010 in some of the eastern states while chronic flood in Bihar often each year has marked influence on National budget.

Table 9.2: Severe Drought and Flood in India.

Year	Drought Area (Per cent)	Year	Flood Area (Per cent)
1877	64.7	1916	32.6
1918	68.7	1933	36.1
1965	42.9	1961	57.1
1972	44.4	1975	40.4
1987	49.2	1983	32.8

In recent year 2009 though, India received only 77 per cent of normal monsoon but heavy rain in Europe in June created havoc in Italy due to the heaviest rainfall and worst flood since last six decades in which 286 people died and 2.5m became homeless. In the same year UK also received 133 per cent rainfall to normal and Ottawa also witness the highest 2434 mm rainfall.

9.1 Uses of Water

Among the different uses of water, 70 per cent of global water is used in agriculture. The uses of water for agricultural, industrial and residential purposes in developed country (US) are in the ratio of 3 : 5 : 1 as against 7 : 2 : 1 ratio of the global uses itself speaks the importance of water in whole world for food sector (Table 9.3).

Table 9.3: Use of Water in Agriculture, Industry and Residential Purposes.

Uses	Global Water Use (per cent)	US Water Use (per cent)
Agriculture	70	33
Industry	20	54
Residential	10	11

9.2 Water Consumption by Crops

There is a linear relationship between crop yield and consumption of water by crop because

1. Carbon and water vapour share a common diffusive pathway between the atmosphere and the interior of plant leaves; consequently stomata must be open and water transpired in order for crops to assimilate carbon from air surrounding leaves.

2. Both water loss and photosynthesis are driven by absorption of light

3. Evaporative water loss significantly cools leaves and canopies, which reduces high-temperature stress.

The very adaptation that favour fast growth in crops-such as leaves with large surface areas, short diffusive pathways from the leaf interior to the atmosphere and high stomatal conductance values also favours water loss.

Changing climatic conditions will affect leaf, plant and crop water-use-efficiency (WUE), is measured and expressed as the relationship between water use and photosynthesis at the crop canopy level and expressed as the total biomass or grain yield of evapotranspiration (ET= Loss of water through plant transpiration + evaporation from plant part and soil).

9.3 Climate Change and Global Hydrology

Technology for forecasting future changes in water resources with changes in global climate and exact hydrological models is yet to available. Though, different countries have their model to predict for changes in precipitation but its probability is beyond perfect. According to one of the prediction enhancement in $[CO_2]$ and temperature may reduce 5 to 10 per cent requirement of water by irrigated crops by 2030 and further 30 to 40 per cent by 2090. Here too, it creates confusion, since CO_2 can only reduce the evapotranpiration by deactivating the stomatal conductance, not the temperature which rather accelerates the ET.

Changes in climate may alter the rainfall pattern as such flood prone area may turn to rain shadow area and drought prone area into an adequate rainfall one. All these changes may bring changes in world ecology. However, some climatologists have forecasted little changes in rainfall patterns than changes in temperature.

9.4 Effect of Elevated [CO_2] and Water Requirement of Crops

Several plants develop their own morphology to thrive well under elevated [CO_2] conditions such as reduction in number and size of stomata and its conductance.

Crops grown in CO_2-enrichment under controlled climate with well mixed air, the stomatal conductance transpiration decreased from 20 to as high as 90 per cent. The enhancement in photosynthesis stimulated the plant growth and produced more yield to per unit water transpired by the leaves. Hence, elevated [CO_2] accounted for higher WUE.

The loss of water may be more because

1. Increasing [CO_2] though decreases leaf transpiration but a proportionate water saving is not realized by whole plant due to feedback operation. Reduced transpiration on a leaf-area basis is counter acted by higher leaf area index (LAI). As the transpiration decreases, the air inter inside and air above the canopy become dryer, increases the driving force evapotranspiration from the canopy leading to further increases in transpiration rate.

2. High CO_2-enriched stomatal closure increases the internal leaf temperature and thus increases the transpiration.

3. The direct effect of CO_2 on leaf-level gas exchange also reduces due to differences in regional- scale vegetation and mixed air.

4. Rising CO_2 promotes the faster stomatal closing and denser crop canopy, interception loss increases at the cost of soil moisture content.

Rising CO_2 may have only a small influence on total water consumption by the crop. Under open field conditions, increases in WUE will be smaller than the predicted one. The WUE of field grown sorghum at high CO_2 was increased by 9 per cent in wet soil and 19 per cent in dry soil but cumulative evapotranspiration was reduced to only 10 per cent in wet soil and 4 per cent in dry soil. In wheat, WUE was enhanced by 21 per cent in high N fed and 10 per cent in low N fed plot but total ET was unchanged by CO_2-enrichment. Similarly sunflower grown in 746 ppm

CO_2, used 11 per cent more water with 26 per cent increase in WUE compared to that grown in ambient CO_2. Thus, these studied indicated that increases in canopy, WUE associated with higher CO_2 are due to increase in rates of photosynthesis but not in less use of water. Crop growth in elevated CO_2 also influences a wide range of plant attributes like carbon assimilation and allocation, growth and nutrient relations to enhance WUE.

In comparison to above ground parts, CO_2-enrichment is more beneficial to roots for their vertical and horizontal growth leading to higher root-to-shoot ratio. This facilitates plant to extract moisture more efficiently to delay wilting under water-stress conditions during prolong drought spell.

It also decreases the hydraulic conductivity along the pathway from roots through shoots and finally into leaves as it was recorded in soybean, sunflower and alfalfa. Under water-logged conditions, CO_2-enrichment develops resistant to water flow in plant, which restricts water absorption and reduces yields but under water-stress conditions, it may help the plant to sustain.

In some other experiments, doubling of CO_2 has negligible effects (10 per cent) on total water consumption by crops. Crop productivity per unit of water consumed is however, likely to increase by 20-30 per cent may be beneficial for dryland agriculture. Though, high CO_2 reduces the water requirement of crops but increases in temperature on the other hand may neutralize the effect at least to some extent but stomatal closing at high [CO_2] may develop plant system to resist ozone pollution.

9.5 Elevated Temperature and Water Requirement of Crops

The extent of transpiration and WUE is influenced by an increase in air temperature, which affects stomatal properties, humidity, leaves morphology and temperature as well as wind velocity. The air moisture in intercellular spaces of leaves is generally very close to saturation irrespective of temperature. As the temperature of canopy increases, the water density of the air within the leaves rises with water density gradient between crop canopy and the unsaturated atmosphere. Rise in temperature results in opening of stomata due to increase in stomatal conductance even at vapour density gradient from leaf to air remained constant, which explains the direct effect of temperature on stomatal opening.

In one of the experiment, the transpiration at 20°C was 4.8 times higher to 5°C and further 2.7 times higher at 35°C compared to 20°C, is an indication that global warming will increase the ET. Rise in temperature is also expected decrease in WUE of both C_3 and C_4 species. Transpiration rate in soybean is increased by almost 4 per cent for each degree increase in temperature. Rice grown at 24-26°C temperature recorded an increase in WUE but it drops to 20 per cent at 30°C. Thus, anyhow warming may influence crop growth and activate development to counteract increased transpiration to some extent.

9.6 Elevated Ozone and Water Requirement of Crops

The effect of O_3 on crop plant is dependent on stomatal conductance. The factors which activate the opening of stomata results in damaging effect while the factors which help in closing of stomata decrease the injury by ozone. Therefore, crops raised in water stress conditions are less susceptible to those raised in presence of adequate moisture due to better opening of stomata. Since CO_2-enrichment also reduces O_3 injury due to reduction in stomatal opening hence, it restricts O_3 penetration into the leaves. Short-term exposure to elevated $[O_3]$ results in closing of stomata while long term exposure results in slow opening of stomata.

Ozone also directly affects metabolism of guard cells particularly lacking cuticular waxes. It uncouples the linkage between guard cells and whole- plant function by interfering with the ABA hormone and Ca^{2+} dependent regulator. Ozone damages the photosynthetic apparatus to retard the carbon fixation and elevates the internal $[CO_2]$ resulting in closure of stomata.

Higher concentration of ozone also interferes with the soil water uptake mechanism and its proper distribution in whole plant. Even the root growth is restricted and population of mycorrhizal and their association with plant roots is weaken leading to loss in soil moisture removal by crops. Reduction in root hydraulic conductivity further reduces uptake of water and nutrients in presence of ozone.

9.7 Species Differences and Water Requirement

Water requirement of different plant species varies as per their physiological requirement as well as from temperate to tropical and arid to semi-arid conditions. In general, the water use efficiency of CAM plants is the highest followed by C_4 species whereas the water requirement of C_3 species is the highest. Among the two major cereals of the world, for each tones of rice production, 2,500 to 3,000 m³ of water is required which is 2.5 to 3.0 times higher to same quantity of wheat production (1,000 m³ water/tone).

Since, C_4 plant's stomata are partially closed in water stress conditions hence, water requirement is less due to inhibition in transpiration which is absent in C_3 plants. The water use efficiency (WUS) is the moles of CO_2 assimilated/moles of H_2O transpired and therefore, WUE of C_4 species is higher than C_3 species due to less respiration.

Transpiration ratio for C_4 plants is in the range of 200-350 as compared to C_3 plants in the range of 500-1,000 itself indicates the higher water requirement of the later.

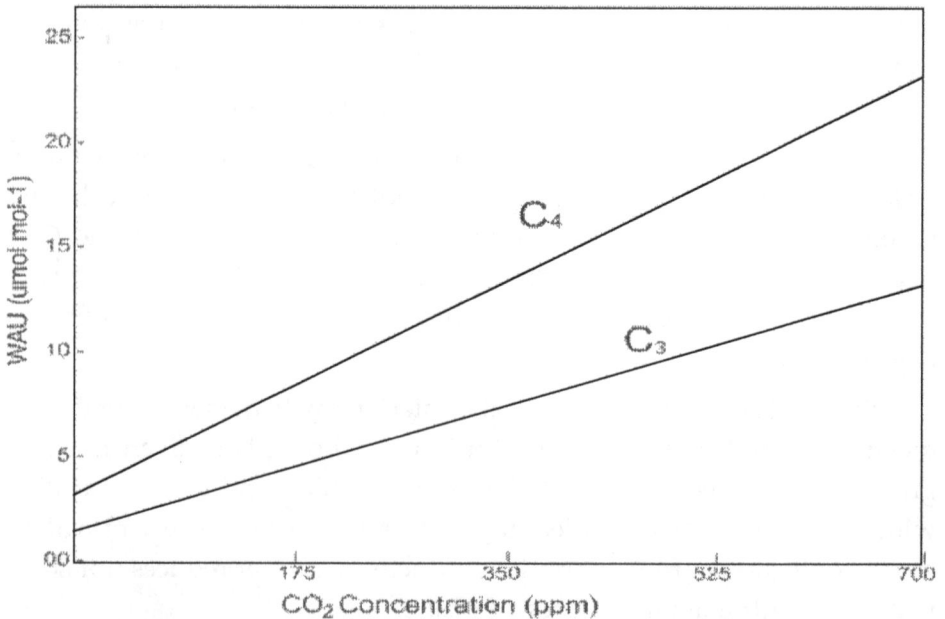

Figure 9.1: Potential Responses of C_3 and C_4 Plants to [CO_2] on Water Use Efficiency (WUE) Indicating C_4 Plants are more Efficient in WUE to CO_2^- enrichment.

Under normal moisture conditions, CO_2 supply limits photosynthesis in C_3 plants to the extent that light saturation occurs at fluence rate about 25 per cent of full sunlight but C_4 photosynthesis never saturate even at full sunlight.

At leaf temperature below 30°C the quantum yield for C_4 plants is less than for C_3 plants. C_4 photosynthesis is less efficient but has higher photosynthetic capacity due to its higher content of components of photosynthetic electron transport and certain photosynthetic enzymes (Rubisco). Therefore, C_4 plant takes advantage at excess available light to generate the ATP needed to run the C_4 cycle, concentrate the CO_2, and increase in net carbon assimilation. However, at low temperature and low light intensity with adequate water supply C_3 perform even in higher CO_2 fixation than C_4 species, as such C_3 species of temperate world has an edge over C_4 in productivity. C_4 plants generally exhibit higher light-saturated rates of photosynthesis than C_3 plants which reflect a higher photosynthetic capacity due to their content of components of photosynthetic transport and certain photosynthetic enzymes (Rubisco). Therefore, C_4 species can use excess available light to generate the ATP needed to run the C_4 cycle, concentrate the CO_2 and increase in carbon fixation.

However, under non-limiting conditions of moisture, lower temperature and radiation, C_3 may be more productive than C_4 such is the case of temperate climates but under high humidity and high temperature conditions C_4 species (Sugarcane, maize and sorghum) are undoubtedly more productive. The typical C_4 plant, bamboo (*Bambusa tulda*) is the most competent to sustain under low to high CO_2 concentration as well as having high capacity to extract moisture and nutrients even from the soils which are constraints in moisture and nutrients.

PART II
Climate Stress

Chapter 10
Population and Pollution

Increase in population and pollution is a simultaneous phenomenon since exploitation of population has direct bearing on the pace of pollution. Wherever *Homo sapiens* inhabited, they polluted the environment and plant as an industry continued to clean the same. The ballooning of the population has virtually abused the natural resources. Thus, over exploitation of the natural resources to meet the demands of the population may be suicidal for life on this bio-planet as cited in the *Rigveda*:

> *"Sky is like Father, Earth is like Mother,*
>
> *And all the creatures that live in Between*
>
> *Constitute a Family. And disturbing to any'*
>
> *One of them will disturb the entire system."*

Even Wordsworth, the great English Poet, has given warning for the consequences of accusing the nature:

> *"Accuse not nature;*
>
> *She hath done her part;*
>
> *Do though but thine;*
>
> *Nature never betray the heart that loved her"*

According to records gleaned from ice cores, tree ring and other "proxy indicators" of temperature, the 1990 was the warmest decade of the second millennium with 1998 the hottest single year. Each of the 12 months from August 1997 through September 1998 set an all-time worldwide high-temperature record with 7 of the 10 warmest years in the past 130 years occurred in 1990s.

During 1998, at least 56 countries suffered severe floods while 45 witnessed droughts with burning of tropical forests from Mexico to Malaysia and from Amazon to Florida. Northern hemisphere received spring a week earlier and freezing periods are also reduced at higher elevations.

10.1 Population and Green House Gases

Climate of the past 1,000 years indicates that human population and its activities are entirely responsible for sharp global warming. Increase in population and industrialization to meet its requirement have simultaneously increased the concentration of greenhouse gases (CO_2, CH_4, N_2O, CFCs) resulting increase in temperature and radiation intensity. Fluctuation in these indices has direct bearing on volcanic activity, oceanic disturbances and several others calamities.

10.2 Population vs. CO_2 Concentration

The increases in population estimated and expected from 1804 to 2060 indicate an abrupt increase in CO_2 concentration after 1970 and onward due to industrialization.

Table 10.1: Growth Rate and Doubling Time of Population.

Region	Growth Rate (per cent)	Doubling Time (yrs)
World	1.2	58
More Developed Countries	0.1	700
Less Developed Countries	1.5	47
Africa	2.5	28
Asia	1.2	58
N-America	0.6	117
Latin America	1.5	47
Europe	− 0.1	−
Oceanic	1.0	70

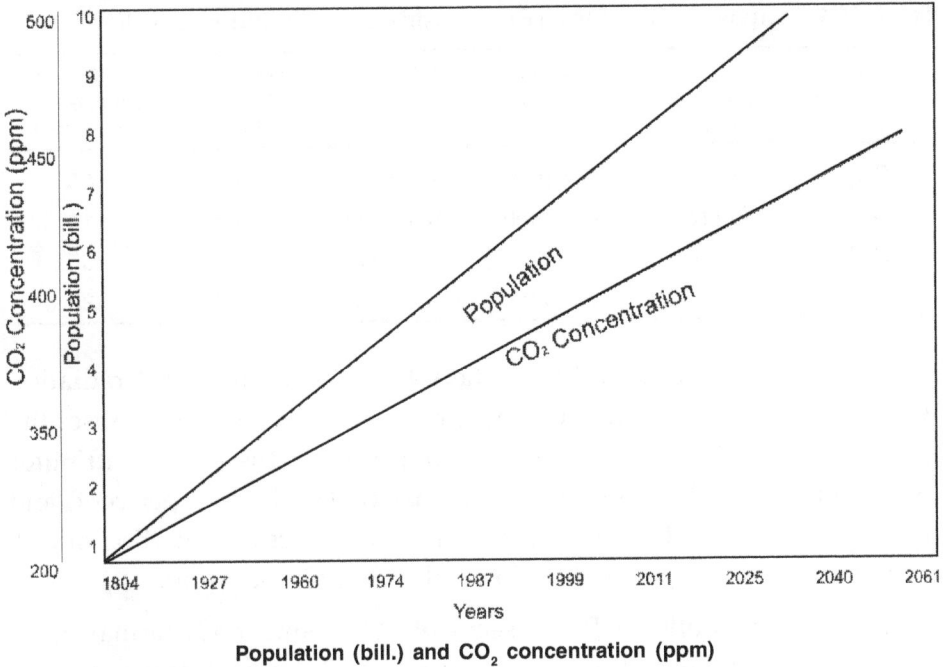

Population (bill.) and CO₂ concentration (ppm)

Figure 10.1: Population and CO_2 Concentration during the Years.

According to another estimate, the Earth's carrying capacity may be from 500 to 1,000 billion people, which is very low to present population of 7.0 billion but its effects are obvious. The growth rate and doubling time of population (Table 10.1) indicates that the world population with a growth rate of 1.2 will swell to double in 58 years. The contribution of African countries will be more with doubling in population only in next 28 years or by the end of year 2038 whereas the contribution of Europe will be at the negative rate (-1 per cent).

The CO_2 concentration of the atmosphere was 230 ppm in 1900 and 325 ppm in 2,000 to 385 ppm in 2007 which is extrapolated to 480 ppm at the end of present century which is much higher beyond the photosynthetic upper limit of the majority of plant species. If, the trends remain in the same pace, it may rise to 740 ppm by 2250. Even the anthropogenic emission of CO_2 which was only 10 ppm during 1850 has multiplied to 45 times during 2010 (450 ppm).

Besides increase in CO_2 concentrations, a marked rise in other green house gases (GHGs) from pre-industrial level to present day level is also worth alarming (Table 10.2).

Table 10.2: Increase in GHGs from Pre-industrial Level to Current Level.

GHGs	Pre-industrial Level	Current Level	Increase since 1750	Radiative Forcing (W/m²)	Contribution (per cent) to Pollution
CO_2	280 pm	387 ppm	107 ppm	1.46	55
CH_4	700 ppb	1745 ppb	1045 ppb	0.48	15
N_2O	270 ppb	314 ppb	44 ppb	0.15	05
CFC- 12	0	533 ppb	533 ppt	0.17	25

These GHGs are in an atmosphere that absorb and emit radiation within the thermal infra-range. This is the main process which causes the greenhouse effects. Among these, water vapour is the main contributor (36-72 per cent) followed by CO_2 (9-26 per cent), CH_4 (4-9 per cent) and ozone (3-7 per cent). However, among the four gases contribution of CO_2 is the maximum (55 per cent) followed by CFC (25 per cent).

Annual contribution of Power sector of GHGs emission is the maximum (21.3 per cent) while industrial emission is 16.8 per cent, transport 14.0 per cent and Agricultural by-product is 12.5 per cent (Table 10.3).

Table 10.3: Annual GHGs Emission by Sectors.

Sl.No.	Sector	Emission (per cent)
1.	Power Station	21.3
2.	Industrial	16.8
3.	Transport	14.0
4.	Agricultural By-products	12.5
5.	Fossil fuel	11.3
6.	Residential and Commercial	10.3
7.	Landuse biomass burning	10.0
8.	Water treatment	3.4

Though, China stands at the top among the 5 major contributors of GHGs but in per capita contribution, US hold the first position which is almost 12 times to what India is producing per capita (Table 10.4).

In respect to production of CO_2 by different countries, the petroleum producing countries are at the top followed by Australia, U S and other industrialized countries while least by the eastern African countries (Table 10.5).

Table 10.4: Top Emitters Countries of GHGs (2005).

Sl.No.	Country	Share (per cent) of Global Total	GHGs/Capita (tones)
1.	China	17	5.8
2.	U S	16	24.1
3.	E U	11	10.6
4.	Indonesia	6	12.9
5.	India	5	2.1

Table 10.5: Major/Minor CO_2 Producing Countries/Capita (t) (2005).

Rank	Country	Tones/capita	Rank	Country	Tones/capita
1	Qatar	55.5	2	UAE	38.8
3	Kuwait	35.0	5	Australia	26.9
6	Bahrain	25.4	7	USA	23.5
11	NZ	18.8	18	Russia	13.7
25	Germany	11.9	36	UK	10.6
37	Japan	10.5	48	South Africa	9.0
74	Brazil	5.4	120	India	1.7
128	Pakistan	1.5	139	Ethiopia	1.0
173	Eritrea	0.1			

Among different continents, North America is at the top in CO_2 production to per capita (24.1 t) whereas Sub-Sahara Africa is at the bottom (Table 10.6).

Table 10.6: CO_2 Production by the Continents (tones/capita) (2000).

Rank	Continents	Tones/capita
1	North America	24.1
2	Europe	10.5
3	Oceania	5.6
4	Middle East and North Africa	5.6
5	South America	5.3
6	Central America and Caribbean	4.5
7	Asia	3.4
8	Sub-Sahara Africa	2.3

The life time of different GHGs, though, concentration (per cent) of Tetrafluoromethane used in deep freezes is very low but its life span is the maximum (50,000 yrs) as compared to CH_4 (12 yrs) and to that of CO_2 is variable due to its utilization by plants (Table 10.7).

Table 10.7: Life Span of Green House Gases.

Gases	Life Span (yrs)	Gases	Life Span (yrs)
CO_2	Variable life time	CH_4	12
N_2O	114	CFC-12	100
CFC-22	12	Tetrafluoromethane	50,000
Hexafluoroethane	10,000	Sulfur-hexafluoride	3200
Nitrogen-trifluoride	740		

Besides CH_4 production, all the industrialized nations are more responsible in polluting the environment since they are emitting more CO_2 and long life span gases including Freon-11 as spray propellant and Freon-12 as coolant in refrigerators, freezers and air conditioners.

Each individual GHG has its different global warming potential (GWP) which is the ratio of heat trapped by one unit mass of the GHG to that of one unit mass of CO_2 (GWP of CO_2 = 1) over specific period of time. The 100 years GWP for NH_4 and N_2O are 23 and 296 times of the GWP of CO_2, respectively while, CO_2 contributes 60 per cent to total global warming followed by CH_4 (20 per cent), CFCs (14 per cent) and N_2O (6 per cent).

10.3 Increases in Global Temperature

Relationship between the oceans and atmosphere indicates that it would easily cause temporary cool or other unpredictable and sometimes sudden changes that would create confusion and paralyze work to affect global warming.

Reading from 616 drilled boreholes on six continents revealed that "subsurface temperatures comprise an independent archive of past surface temperature changes that is complementary to both the instrumental record and inmate proxies, such as tree-ring. Some 80 per cent of that warming corresponds with the growth of industrialization. Borehole temperatures indicated a greater warming over the past 5,000 years than temperatures computed by other means.

According to one of the observation in 2009, the world sea level raised to 20 cm since 1900. The temperature of the Antarctic Sea has also increased by 0.1°C and world atmospheric temperature by 0.6°C since start of the 20th century. The world atmosphere is getting heated since 1940 as rising above 0°C to 0.6°C during 2000 (Table 10.8).It appears that the atmospheric temperature is rising by 0.1°C in each decade.

Table 10.8: Increase in Temperature Since 1900 (0°C).

Year	Temperature (0°C)
1900	– 0.15
1950	+ 0.10
2000	+ 0.60

This increase in atmospheric temperature is resulting reduction in the height and spread in glacial and thus increases in Sea-level is causing danger of submerging to some of the low land islands. Recent report has warned that the melting Arctic is now being called an "**economic time bomb**". Economic modeling shows methane emissions caused by shrinking seas ice from just one area of the Arctic could come with a global price tag of $69 trillion- the size of the world economy in 2012.

Scientists of Cambridge and Rotterdam calculated the potential economic impact that methane from the East Siberian Sea will be emitted as a result of a thaw. This constitutes just a fraction of the vast reservoirs of methane in the Arctic, but researchers believe the release of even a small proportion of these reserves could trigger possibly catatrophic climate change.

Gail Whiteman, University of Erasmus, Rotterdam, Netherlands termed it as the "economic time-bomb," due to global impact of a warming Arctic. The massive methane boost will have major implications for global economies and societies. It also explored the impact of a number of long-lasting or smaller pulses of methane and in all these happening, the economic cost of physical changes to the Arctic is "steep".

10.4 Increases in Solar Irradiance (W $^{-2}$m)

Increases in CO_2 and other GHGs concentrations and simultaneously decreases in O_2 concentration has depleted the ozone layer. Thinness or hole in ozone layer allows the high intensity shorter wave length radiation to pass and travel to troposphere resulting heating of the earth and its atmosphere.

Table 10.9: Increases in Solar Irradiance ($W^{-2}m$).

Year	Solar Irradiance ($W^{-2}m$)
1900	1365.5
1950	1367.5
2000	1368.0

A marked rise in solar irradiance in past 100 years has been recorded (Table 10.9) If, this trend is continued, may change the ecology of the green planet.

Chapter 11

Soil and Water Pollution

Presence of toxic substances both in soils and water are equally injurious for plant life in the same way since the soil pollutants play their roles only when these are dissolved into water and interfere in absorption of water and nutrients. Presence of heavy metals in soils or through industrial waste products coming directly through air or streams as well as deposition through urban sewage on one hand and application of toxic chemicals as pesticides, herbicides and fertilizers on the other are the main sources of soils and water pollution.

11.1 Causes of Soil Pollution

☆ Runoff from pollutants (paint, chemicals rotting organic materials) leaching out of landfill.

☆ Polluted water discharged from factories.

☆ Oil and petroleum leaks from vehicles washed off the road by the rain into the surrounding habitat.

☆ Chemical fertilizer runoff from farms and crops.

☆ Acid rain (fumes from factories mixing with rain).

☆ Sewage discharged into rivers instead of being treated properly.

☆ Over application of pesticides and fertilizers.

☆ Purposeful injection into groundwater as a disposal method.

☆ Interconnections between aquifers during drilling (poor technique).

☆ Septic tank seepage.

☆ Lagoon seepage.

☆ Sanitary/hazardous landfill seepage.

☆ Cemeteries.

☆ Scrap yards (waste oil and chemical drainage).

☆ Leaks from sanitary sewers.

11.2 Effect of Soil Pollution

☆ Pollution runs off into rivers and kills the fish, plants and other aquatic life.

☆ Crops and forages grown on polluted soil may pass the pollutants on to the consumers.

☆ Polluted soil may delay the maturity of crops and forages.

☆ Soil structure is damaged (clay ionic structure impaired).

☆ Corrosion of foundations and pipelines.

☆ Impairs soil stability.

☆ May release vapours and hydrocarbon into buildings and cellars.

☆ May create toxic dusts.

☆ May poison children playing in the area.

11.3 Soil Pollution Effect on Plants

Atmospheric pollutants are settled on ground surface as well as deposited on the soil as sewage, industrial effluents and agricultural wastes effects increases soil acidification.

The oxides of sulphure and nitrogen, chlorides, fluorides, ammonium etc. emitted into the atmosphere in combustion from various industries come down as dry or wet deposition (acid rain) onto the soil and lower the soil pH. Increased acidity of soil results in following effects.

The activity of soil microbes, particularly of decomposers, is reduced. The decomposition of organic matter and consequently nutrient cycling in the soil is reduced. It ultimately adversely affects the growth of plants.

The bases in the soil are leached down due to soil acidity. As exchangeable bases become deficient in the soil, plant growth is reduced due to nutrient deficiency.

The roots, particularly the root hairs, are damaged resulting in reduction in nutrient uptake by plants.

Increased acidity mobilizes heavy metals like Al, Cd, Zn, Hg, Mn, Fe etc. These spread rapidly in the soil along with soil water and reach concentrations toxic to plants. Consequently, plants show species specific metal toxicity symptoms. Al-toxicity generally damages root hairs and reduces nutrient uptake while Fe-toxicity has general adverse effect on plant growth. In some soils, acidification increases weathering of silicate minerals destroying the mineral structure of the soil. This leads to poor growth of vegetation in general.

In some marginal soils and grasslands, acidification increases the supply of plant nutrients like sulphur and nitrogen. The vegetation is thus, benefited by soil acidification and plants may show better growth.

11.4 Effect of Pesticides

Various pesticides, insecticides, fungicides etc. are used in agriculture as foliar spray or are applied to soil far in excess to the requirement causing soil pollution. These substances pollute the soil depending upon their volatility, biodegradability, persistence, leaching, chemical reactivity and adsorption on the soil particles. Many of these substances form cations, are adsorbed on silicate clay micelle or humus molecules on the pH-dependent exchangeable charge sites and are later absorbed by the plants. Absorbed pesticides substances produce characteristic species-specific toxicity symptoms in plants just like their aerial overdose. Fungicides reduce abundance of soil fungi and actinomycetes and interfere with decomposition of soil organic matter adversely affecting the nutrient cycling.

Pesticides increase the abundance of some bacterial species, particularly of ammonifying bacteria while reduce the abundance of some susceptible bacteria.

Insecticides reduce the abundance of predator soil microbes and consequently increase the abundance of their prey species. In general, species composition of the soil microflora and fauna is changed by pesticide substances.

The inorganic pesticides contain arsenic and sulphur. These substances cause toxicity symptoms like yellowing, necrosis, shot holes, premature defoliation etc. in plants. Among the different pesticides, di-eldrin and silver chloride are more toxic than methyl parathion and endosulfan. Chlorine gas pollutant decreases germination in wheat and clovers and further causes leaf necrosis, marked deformation of epidermal cells and hairs. Mutagenic effects of factory pollutants on these two crops also occur through appearance of **novel bands**.

11.5 Effect of Herbicides

Various herbicides are used for weed control in agricultural practice. General effects of some common herbicide substances on the plants are as follows.

The symptoms in response to a particular herbicide may be characteristic but their development depends upon the dose to which plants are exposed, rate of growth of plant, weather conditions and the plant species. Acetanilides (*e.g.* Prochlor, Metachlor) cause stunting of plants and roots. In brassicas, these cause yellow, red and blue colouration of leaves with pronounced stunting.

Amides and carbamates (*e.g.* Diphenamide, Propyzamide, Chlorpropham and Asulam) cause stunting without chlorosis or leaf-scorch in mild doses. Crops generally become greener in colour. Roots show thickening and stunting. In the cereals, coleptile becomes stunted and swollen.

Benzonitriles (*e.g.* Dichlorobenil) cause stunting of plants and thickening of roots. In shrubs and bushes, bark at the ground level develops necrosis. In some cases, marginal yellowing occurs in the leaves.

Aliphatic acids (*e.g.* TCA, Dolpon) cause stunted growth of shoot, loss of leaf wax and mild irregular chlorosis.

Growth regulators (*e.g.* 2,4 D, MCPA, Chlopyralid, and Mecoprop) cause hormone-type distortion in leaves and fruits. In the brassicas, stem splitting occurs. In the cereals, ears become distorted, becoming blind or with shriveled grains.

Ureas and Uracils (*e.g.* Diuron, Linuron, Bromocil and Lencil) cause inhibition of photosynthesis leading to yellowing along the veins that later extends across whole leaf. In the cereals, chlorosis starts in the middle of leaf but extends quickly to the tip.

Sulphonyl ureas (*e.g.* Chlorsulfuron) cause inhibition of growth in susceptible plants and these ultimately die due to root growth being stopped.

Triazinones (*e.g.* Metribuzin) cause inhibition of photosynthesis that leads to yellowing and tip-scorch in the leaves. Triazines (*e.g.* Atrazine, Simazine) cause chlorosis and necrosis spreading towards tips of leaves and inhibit photosynthesis. This is more severe in older leaves.

Thiocarbamates and ethofumesates cause chlorosis, stunting and leaf-curling. Younger leaves become stuck to older leaves. In the cereals, this condition often leads to characteristic pink colouration at the base of the stem.

11.6 Effect of Sewage and Ash Pollution

Sewage matter is commonly used as fertilizer or deposited as waste on the soil. Effects of such pollution are mostly common to all plants.

The organic matter in sewage decomposes and produces nitrogenous substances that become excessive in the soil and harm the vegetation.

Decomposition of sewage may also release various toxic heavy metals that cause characteristic heavy metal toxicity symptoms in plants.

Detergent substances may also be released from sewage causing characteristic injury to plants.

Ash produced mainly from combustion of coal in thermal and industrial plants used for land filling or deposited on soil makes the soil unfit for vegetation. It may release many toxic substances in the soil causing characteristic plant injuries.

11.7 Effect of Fertilizers

Fertilizers are generally used far in excess of the requirements of the crop. The unutilized fertilizers cause soil pollution. Toxic concentrations of nitrogen fertilizers cause characteristic symptoms of nitrite or nitrate toxicity in plants, particularly in the leaves. Nitrogenous fertilizers generally cause deficiency of potassium, increased carbohydrate storage and reduced proteins, alteration in amino acid balance and consequently change in the quality of proteins.

Ammonium fertilizers produce ammonia around the roots that may escape in the soil and cause ammonia injury to plants. Ammonium and nitrate produce acids in the soil and increase soil acidity.

Nitrate and nitrite bacteria are reduced while ammonifying bacteria are increased in the soil disturbing the nitrogen cycle. Excessive potash in the soil decreases ascorbic acid and carotene in the plants. Superphosphates cause deficiency of Cu and Zn in plants by interfering with their uptake.

Excessive lime prevents the release of Co, Ni, Mn and Zn from the soil and their uptake by plants is reduced causing their deficiency symptoms.

Excessive deposition of various substances released from chemical fertilizers into the soil generally causes their over-absorption by plants. These over-absorbed substances become accumulated in plant parts (bioaccumulation) *e.g.* nitrogen and sulphur are deposited in the leaves.

11.8 Effect of Industrial Effluents

Various inorganic and organic substances are present in the industrial effluents. These substances mostly remain tied up in the soil and are not readily available to plants. However, they affect various soil characteristics.

Effluents affect the mineral structure, soil pH, exchangeable base status etc. of the soil and thus indirectly affect the plants since pH of the soil is disturbed making soil either acidic or alkaline.

Various inorganic and organic chemicals are accumulated in the soil up to levels toxic to plants. In highly polluted soils, plants absorb and accumulate toxic substances (bioaccumulation). These substances may or may not produce direct injury symptoms in plants but are passed on to higher trophic levels (biomagnification).

11.9 Effect of Radioactive Pollutants

A variety of radioactive waste materials like Strontium-90, Cesium-137, Iodine-131, Plutonium, Uranium, Americium, Curium, Neptunium etc. are added to soil from nuclear activities. These substances may be washed into water or may be directly added to water that is used as coolant in nuclear power plants.

Uptake of radioactive substances by plants depends upon pH and organic matter status of the soil.

Various radioactive materials show different solubility and absorption by plants. Plant uptake is generally highest for Neptunium, lowest for Plutonium and intermediate for Americium and Curium. Strontium-90 behaves as Ca in the soil and is absorbed by plants like it. Cesium-137 behaves like K. but its uptake by plants is very limited.

Absorbed radioactive substances are generally accumulated in the vegetative parts of the plants, particularly the leaves. Their accumulation in crop fruits and seeds is very low.

Accumulation of absorbed radioactive substances in the plants may be up to 100 times to their concentration in the soil and water. Radiation from radioactive substances may also adversely affect the plants. Pines are eliminated on exposure to 20-30 roentgen/day while most plants die at 200roentgen/day. Only lichens and mosses are highly resistant to radiation. Radiation also damages chromosomes. It increases the frequency of chromosomal aberrations and causes genetic mutations. Such genetic changes may adversely affect plant metabolism or change their characteristics in subsequent generations.

11.10 Causes of Water Pollution

11.10.1 Toxic Wastes Production by Industries

Filtering of heavy metals into water are fatal to aquatic life and such fishes in menu are also toxic for human consumption since it is linked with birth defects, fertility problems and even cancer. Contamination of water with cement, lubricants, plastics and metals by the construction industries is the other sources of water pollution.

11.10.2 Agriculture Sector

Application of insecticides, pesticides, weedicides and even some fertilizers by the farmers led to seeping into the ground water or run off into ponds, lakes and rivers is among the major contributor to water contamination. Ground water pollution also occurs when chemicals, debris, garbage, oil or other harmful contaminants enter the ground water supply over time. The same water is pumped out and if, taken without proper treatments may be a health hazard.

Natural catastrophes such as heavy storms, earthquakes, floods, volcano eruptions and acid rains disrupt the ecological balance and pollute the water.

The process of contamination begins with tiny organisms called zooplankton that travels up the food chain through clams, birds, marine creatures, and ultimately, humans. These types of algae and considered toxic because they have been known to cause fatalities in humans. Even overgrowths of non- toxic types of algae can effectively block the sunlight from penetrating the water's surface, which makes it difficult for marine life to find food, causing eventual death.

11.10.3 Effect of Water Pollution

The effects of water pollution are numerous (as seen above). Some water pollution effects are recognized immediately, whereas other did not show up for months or years. Additional effects of water pollution include the followings:

11.10.4 The Food Chain Damage

Toxins travel from the water to the animal that drinks and passes to humans when the animal products in terms of meat and milk are consumed.

Diseases can spread via polluted water. Infectious diseases such as typhoid and cholera can be contracted from drinking contaminated water. This is called microbial water pollution. The human heart and kidneys can be adversely affected if polluted water is consumed regularly. Other health problems associated with polluted water are poor blood circulation, skin lesions, vomiting, and damage to the nervous system. In fact the effects of water pollution are said to be the leading cause of death for humans across the globe.

Acid rain contains sulfate particles, which can harm fish or plant life in lakes and rivers. Pollutants in the water may alter the overall chemistry of the water, causing changes in acidity, temperature and conductivity. These factors have an adverse effect on the marine life. Marine food sources are contaminated or eliminated by water pollution. It altered the water temperatures (due to human actions) can kill the marine life and affect the delicate ecological balance in bodies of water, especially lakes and rivers.

11.11 Pollution and Heavy Metals in Soil-Plant-Animal Chain

A metal having specific gravity greater than 5 or atomic number greater than 20 is termed as a heavy metal. Sometime a metal heavier than Ca is also known as heavy metal. Ag, Au, Cd, Cr, Hg, Mn, Pb, Sb, Sn, Te, W and Zn are the most hazardous metals in the biosphere, while Be, Cd, Cr, Cu, Hg, Ni, Pb, Se, V and Zn elements have greater risk to environmental health. Joint Expert Committee of FAO/WHO declare, Cd, Pb, Hg and As as the most toxic heavy metals for human health while World mining of elements with their production in order is as Iron, Aluminum, Copper, Manganese, Zinc, Fluorine, Chromium, Titanium, Barium, Lead, Boron, Zirconium, Nickel, Bromine, Tin, Molybdenum, Uranium whereas Silver stands at 26th and Gold at 32nd. Soil Factors Affecting Heavy Metal accumulation in plants:

The following soil conditions are favourable for heavy metal accumulation in plants.

1. Availability of Pb and Cd decreases with increasing pH.
2. Availability of Pb and Cd also decreases with increasing CEC.
3. Binding of Pb and Cd is closely related to organic matter content of soils.
4. Soil moisture stress conditions are favourable for availability of heavy metals due to stronger cations as compared to weaker essential nutrients.
5. Areas nearer to mines and industries are more prone to soil contamination and their accumulation in plants.

Therefore, plants growing in low pH soils under low organic matter and moisture stress conditions followed by highlight intensity and high temperature in general and close to mines and industrial areas in particular have every possibility for heavy metals toxicity.

11.11.1 Action of Heavy Metals

Heavy metals are "Lewis acids", which can accept a pair of electron from a coordinate co-valent bond; that is, they react with naturally occurring "Lewis bases" in the cell such as $-S^-$, $- OH^-$, amino groups as well as carbolic acid termini. Cd, Pb and Hg affect by activating sulfhydral groups and N atoms in protein. For a redox-active metal an excess supply results in uncontrolled redox reactions, giving rise in formation of toxic free radicals as:

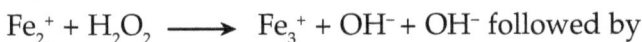

$$Fe_2^+ + H_2O_2 \longrightarrow Fe_3^+ + OH^- + OH^- \text{ followed by}$$

$$Fe_3^+ + H_2O_2 \longrightarrow Fe_2^+ + O.\,OH + H^+$$

These free radicals may lead to lipid peroxidation and membrane leakage.

Cd affects photosynthesis, Calvin-cycle primary process which leads to "down regulation" of PSII. It reduces mineral concentrations of Mn, Cu and chlorophyll in the leaves (*Brassica juncia*).

Cd-resistance is associated with presence of SH- containing phytochelation (PCs) which bind the metal and restricts the entry into the root.

11.11.2 Genotypic Differences in Heavy Metals Uptake

The concentration of Ni seems to be very high in vegetative parts of a number of species from 9 ppm in oats to 21ppm in wheat. Cd concentration in vegetative parts on dry weight basis of different crop species grown on a soil enriched with 10 ppm Cd by sewage sludge in their vegetative parts (Dry matter) were the highest in

Turnip (160ppm) and the lowest in upland rice (1 ppm) and low land rice (0 ppm).while in Grain/fruits/roots (Fresh weight) were the highest in carrot (20 ppm) and the lowest in upland rice (2 ppm). In general, heavy metals content in different parts of the vegetable plants are in order of leaves > roots/tubers > fruits.

11.11.3 Heavy Metals Tolerance

Uptake and phyto-toxicity of Cd, as compared to Pb, Ni, Hg, or Cr is higher for most of the crop species. Cd toxicity is often correlated with Fe - chlorosis. Among crops, soybean is the most susceptible to Cd toxicity, while rice is the most resistant.

Cadmium

Since, Cd as being one of the most eco-toxic metal in the environment, has highly adverse effects on soil biological activity, plant metabolism and human/animal health hazards hence, it is growing worldwide environmental concern. In acid soils within the range of pH 4.4 - 5.5, Cd is the most mobile element, where as in alkaline soils, it is rather immobile. 95 per cent of Cd sorption takes place within 10 minutes, reaching equilibrium in one hour and the soil has very high affinity for Cd at pH 6.

In the pH interval from 4.0 to 7.7, the sorption capacity of the soil increases about 3 times for a pH increase of one unit.

The linear relationship between Cd in plant and in growth medium while, higher concentration of Cd in crops grown under dry soil conditions as compared to saturated conditions were recorded

On dry weight basis, Cd concentration in cereal grains ranges from 0.013 to 0.22 ppm, in grasses from 0.07 to 0.27 ppm and in legumes from 0.08 to 0.28ppm. The safe limit of Cd in major cereal (wheat) is 0.05 ppm. Rice is the most resistant crop to Cd toxicity, while soybean is the most susceptible. Edible part of carrot (19ppm) and leaves of tomato (125ppm) contain the highest concentration of Cd.

Interactions with Other Elements

   ☆ Cd x Zn - Application of Zn reduces Cd uptake.

   ☆ Cd x Ca - Application of Ca also reduces the availability of Cd.

   ☆ Cd x P – Phosphorus is synergistic to Cd hence; application of phosphatic fertilizers increases the Cd concentration in soils.

   ☆ Cd x Cu and Cd x Fe interactions are also very common.

Effect on Human Health

The "classical" example of Cd toxicity was the occurrence of human sickness *Itai – Itai* disease in Japan, caused by Cd released from industrial areas and passed from soil into the food chain via plants. Only a part of Cd in topsoil is immediately available to plants. Consumption of such plants may cause an increase in Cd concentration in kidney and over 300 ppm concentration interferes with its normal function.

Mercury

Effect on Plants

Almost a linear response to increase in concentration of Hg on roots and shoots of 7 days old seedlings of oats is found. There is a little additional uptake of Hg by plants even grown on excess Hg containing soils observed. Hg vapour is influenced by illumination but not by ambient temperature. Even 1 ppb Hg concentration in nutrient solution is injurious. Stunting of seedling growth, root development and inhibition of photosynthesis are the common symptoms of Hg toxicity. It also antagonizes the K uptake.

Concentrations in Plants

In vegetables, Hg concentration varies from 2.6 to 86ppb and in fruits it ranges from 0.6 to 70ppb on dry weight basis (DW). The concentration in grasses and legumes goes up to 100ppb (DW), while it is low in cereals (0.9 to 21ppb). Plants growing in mining or industrial areas in general and particularly lichens, lettuce, carrots and mushrooms contain higher concentration of Hg (72 to 200ppm on dry weight basis).

Effect on Human Health

Serious risk is involved in consuming crops grown in mining areas of heavy metals since the maximum health hazards from Hg and Pb metals are recorded.

Lead

Effect on Plants

Pb is naturally available in plants but it is neither an essential nor has any metabolic role in plant. Pb absorption is passive and the rate of absorption is reduced by liming in acid soils and low temperature also reduces its availability. Higher concentration of Pb in plant's parts of larch (*Larix dahuria*) with increase in Pb concentration in soil has been recorded.

Almost linear response to Pb nutrition on plants was recorded as the days were advanced from 7 to 60. Higher concentration (200 mg/L) of Pb in solution also recorded higher concentration of the element in underground and above ground parts of the plants as compared to lower concentration (25mg/L). Concentration of Pb was higher in roots than in shoots. Airborne Pb, is the major source of Pb absorption by plant foliage. Pb deposited on the leaf surface is absorbed since Pb is fixed to hairy or gummy cuticle of the leaves, which contributes about 95 per cent of the total plant content.

Interaction with other Element

Pb is synergistic to Cd uptake but antagonistic to Ca. Liming reduces Pb activity and induces the availability of Ca and P. Pb is also unfavourable to S absorption.

Concentrations in Plants

Plants grown in uncontaminated soils usually contain an average of 2ppm Pb on dry weight basis. The proposed safe levels for Pb in foods are as: Cereals, 0.2 ppm; potatoes, 0.l ppm, FW; forage grasses, 2.1 ppm, DW and legumes, 2.5 ppm.

Effect on Human Health

The effect of anthropogenic Pb in soils and its effects on human health are very hazardous, which comes through food chain and soil dust inhalation. If its level in soils increases beyond 500 ppm, it is injurious for plant as well as for man and animal.

Arsenic

Effect on Plants

Drastic reduction in corn growth was recorded when concentration of arsenic in soil was increased from 1 to l000ppm. The adverse effect of element on crop grown in heavy soil with high organic matter was more pronounced as compared to crop grown in light soil having low organic matter.

Concentration in Plants

Arsenic concentrations in plants grown in uncontaminated soils vary from 0.009 to 1.5ppm on dry weight basis with leafy vegetables in higher range and fruit in lower range. Grasses contain from 1.1 to 5.4 ppm of As while mushrooms from 10 to 38 ppm. Plants grown in mines and industrial areas may contain more than 6000ppm of As on dry weight basis Sea vegetation also contain higher As concentration.

Nickel

Effect on Plants

Nickel is required by legumes for nodulation as a substitute for Mo or Co It also helps in mineralization of N as the concentration of Ni in soil solution is increased, the uptake rate of Ni by soybean gave a linear response and its concentration in tobacco leaves was also recorded the same trend.

Ni in Solution

The fast availability of Ni by roots, when it is in soluble phase, up take is positively correlated with its concentration in solution.

Interactions with other Elements

Presence of Cu, Zn and Fe inhibits Ni absorption. Excess of Ni causes Fe deficiency by restricting the translocation of Fe from root to shoot.

Concentrations in Plants

On an average grasses contain from 0.1 to 1.7 ppm, clovers from 1.2 to 2.7 ppm, vegetables, 0.2 to 3.7ppm and wheat, 0.2 to 0.6 ppm of Ni on dry weight basis. Ni concentration in cereals grain is less than in straw. Airborne and sewage sludgs are the main sources of Ni pollution and health hazards.

In addition to heavy metals toxicity on human health directly through water or indirectly via plants several other diseases like cardiovascular and respiratory due to air pollution have caused 150,000 lives/annually during past 30 years Temperate latitudes, Pacific and Indian Ocean are happened to be the potential vunerable regions. This has also resulted in malnutrition as an after effect due to occurances in El Nino, Southern Oscillattion and drought in Sub-Sahara African countries. Frequent spreading of Gastroenterological diseases in half fed population of sprawling cities due to water pollution or even shortage of drinking water is quite common.

Chapter 12
Light Energy Harvest

Doubling in world human population by the year 2058 to the 2,000 level will be a challenging task for the scientists and farmers to feed them. Since expansion of the land cannot be done hence we have to meet the demands for food, housing, roads, industries and several other establishments from the same land. About 14Gha(x10⁹) of land in the world, only 1.4Gha are arable non-stressed crop land. Of the remaining area, 2.9Gha are under mineral stress including 1.0Gha of salt stress, 3.7Gha endure drought stress, 1.6Gha are limited by excess water, 3.2Gha have shallow soil profiles and 2.0Gha are subject to permanent freezing. The total arable land potentially arable land, which could produce reasonable crop harvest, is estimated at about 3.2Gha [Christiansen, 1982, (CAB International, 2000) eds. K R Reddy and Hodges]. If, it is computed to per capita, it is 4720m²/capita. The situation of India is more alarming where only 1400m²/capita arable land is available.

According to another estimate, an average human body requires about 100k cal/hr energy which is almost equivalent to the electrical energy consumed by a 100 Watt bulb/hr. Even the radiant energy received to per m² in UK (105W/hr/m²) is sufficient to meet the energy requirement of one person, if, it is harvested with 100 per cent efficiency by the crop plants. This value is very high for other tropical countries like US (185W/

hr/m^2), Australia (200W/hr/m^2) and Red Sea area (300W/hr/m^2). This value may be somewhere between 200 to 300W/hr/m^2 for India.

Though, in these days climate change is of more concern but global change is of greater concern than the former. The phrase of Earth system refers to the interacting physical, chemical, biological and human processes that transport and transform materials and energy and thus provide the conditions necessary for life on this planet. Climate refers to aggregation of components of weather- precipitation, temperature, cloudiness, for example – but the climate system includes processes involving ocean, land sea ice in addition to the atmosphere. The Earth system changes, natural or human driven that can have significant consequences without any changes in climate. The global change should not be confused with climate change- it is more than climate change. Since impact of climate change is of greater importance in crop production hence, it is obvious to discuss the component of climate change of which solar radiation is one of them to directly affect the agricultural world.

12.1 Photo-synthetically Active Radiation (PAR)

The solar radiation on the basis of light wavelength (Figure 12.1), is of seven types ranging from cosmic light (10^{-12}) to radio light (10^3).

PAR= 4x10^{-5} – 7x10^{-5}						
10^{-12}	10^{-10}	10^{-7}	10^{-6}	PAR	10^{-3}	10^3
(cm)	Cosmic	Gamma	X-ray	UV	IR	Radio

Figure 12.1: Electrogenic Radiation, Photo-synthetically Active Radiation (PAR) Amounts for Only a Very Narrow Waveband.

The detail of PAR or visible light is further illustrated (Figure 12.2). Light wave length between 400-500 ηm is a combination of violet, blue-green while between 500-600 ηm is consisted of green, yellow green, yellow and orange colours and relatively a broader wavelength of red spectrum (600-700 ηm).

Radiation of wavelength, 320 ηm is known as UV-radiation which is further divided into UV-A (320-700 ηm), UV-B (290-320 ηm) and UV-C (200-290 ηm) radiations. Radiation between 400-700 ηm amounts for only

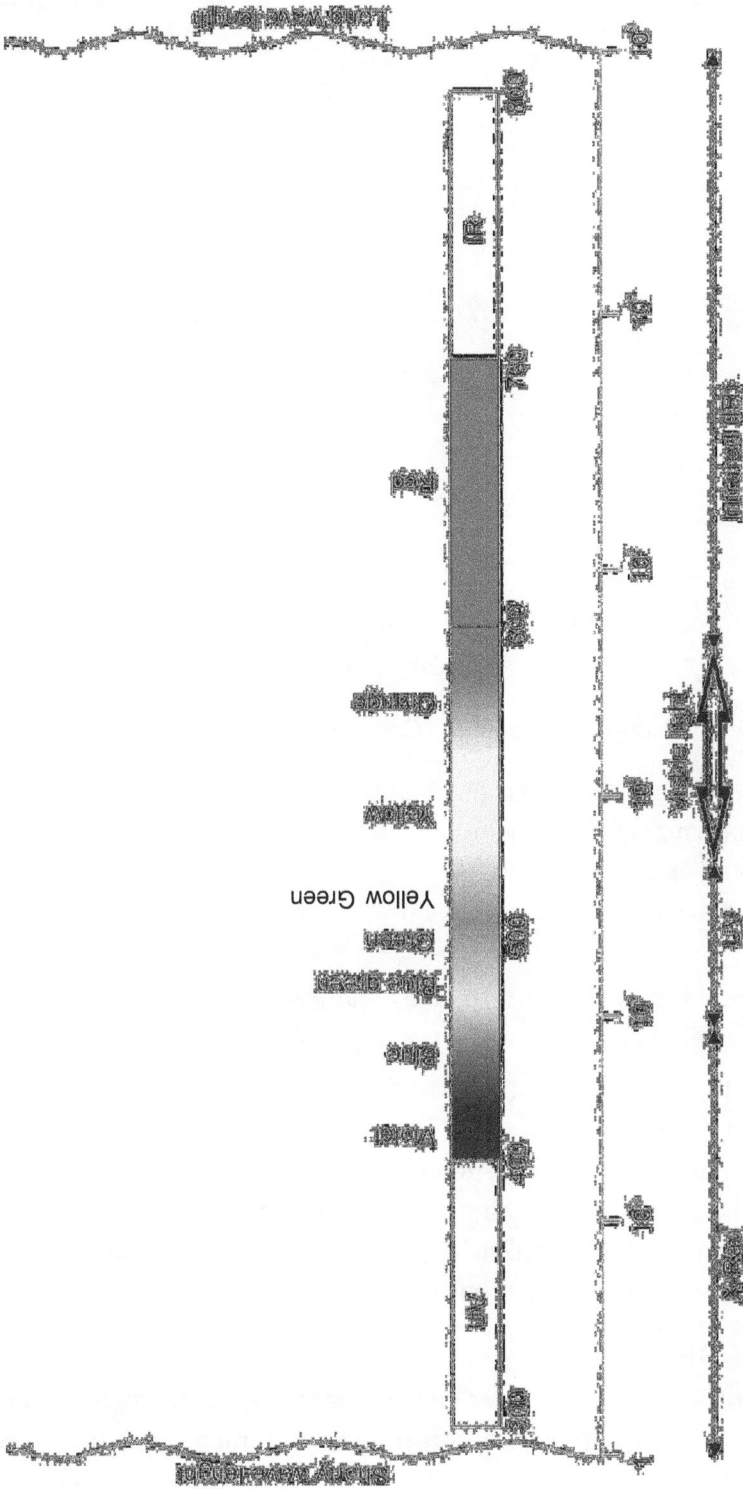

Figure 12.1: Wavelength (λ) and Other Aspects of Wave Nature of Light.

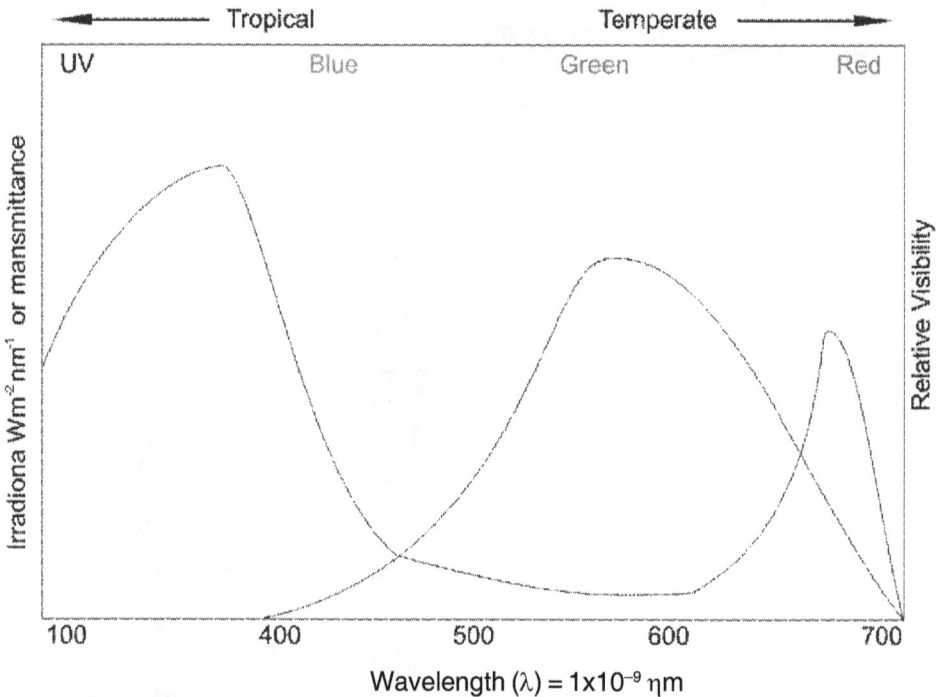

Figure 12.2: What Man Sees and What Green Plant Needs.

Wavelength between 400-700 ηm (PAR or Visible light)
400 ηm (upper limit of UV-radiation)
700 ηm (lower limit of the near infrared)
555 ηm (for the optimum vision of human)

a very narrow wavelength is termed as photo-synthetically active radiation (PAR) or visible light (Figure 12.2).

The optimum photosynthesis in tropical plant species occurs at light wavelength around 400 nm while in temperate plant species it happen at wavelength around 700 nm. Therefore, UV-B radiation which is coming in the troposphere due to thinness of ozone layer is the problem to be faced by plant species as well as for human health. The colour, wavelength and energy value of the different radiations is further listed (Table 12.1).

12.2 Maximum Yield

If a crop can convert 5 per cent of light energy to chemical energy then the maximum yield based on the mean annual irradiance (*i.e.*, the

Table 12.1: Types of Radiation and their Average Energy.

Colour	Wavelength (nm)	Av. Energy (kJ/mole proton)
Ultraviolet	100-400	47
UV-C	100-280	399
UV-B	280-320	332
UV-A	320-400	–
PAR/Visible	400-740	–
Violet	400-425	290
Blue	425-490	274
Green	490-550	230
Yellow	550-585	212
Orange	585-640	196
Red	640-700	181
Far Red	700-740	166
Infrared	> 740	85

light energy average over 24 hours and 365 days), the most efficient sorghum (*Pennisetum typhoide*) plant can harvest 4.25kcal/g dry weight. For UK, an annual irradiance of 105W/m² (equivalent to 25calm⁻²s⁻² or 90k cal.m⁻²h⁻¹), 5 per cent of this value (4.5kcal) would give a dry weight equivalent to over 1gm⁻²h⁻¹ (about 9 kg m⁻² year⁻¹ or, 90t/ha). Though, this estimate may be higher for the countries, receiving higher light energy (>105W/m²) but under field conditions it hardly occurs since only a part of ground surface is covered by the plant canopy. Even if the soil surface is fully covered, solar radiation is not the only factor towards yield performance. The duration of the crop its photo active period, temperature, soil moisture, nutrient supply, pest control and several other factors are responsible.

12.3 Structure of Atmosphere

The process of global warming or green house effect and position of incoming solar radiations, gases and temperature and role of ozone layer can be well understood from the structure of atmosphere (Figure 12.3). From Sea level to 500 km, the atmosphere is divided into four spheres.

Structural of Almosphere

Incoming Solar Radiation

500km

Thermosphere

85km

Mesosphere

50km

Stratosphere

10-16km

Troposphere

Sea level

Infrared visible & ultraviolet >330nm

Penetrate to earth surface

Ultraviolet > 220 < 330 nm

Penetrate to 50 km from sea level and break the ozone layer

Higher energy ultraviolet 100 nm

Penetrate to 200 km from sea level

$[o] >> [o_2]$

o_2^+, o^+ No$^+$

1200^0C

120km; $[o] - [o_2]$

O$^+$ NO$^+$
-92°C

* * * * * * * * *
* * * * * * * *
* * * * * * * Ozone layer

O_3 +hv (220-330nm) =O_2+O

N_2 , O_2, Co$_2$, H$_2$O

Cloud

* * * * * * * * *
* * * * * * * *
* * * * * * *

Land Surface

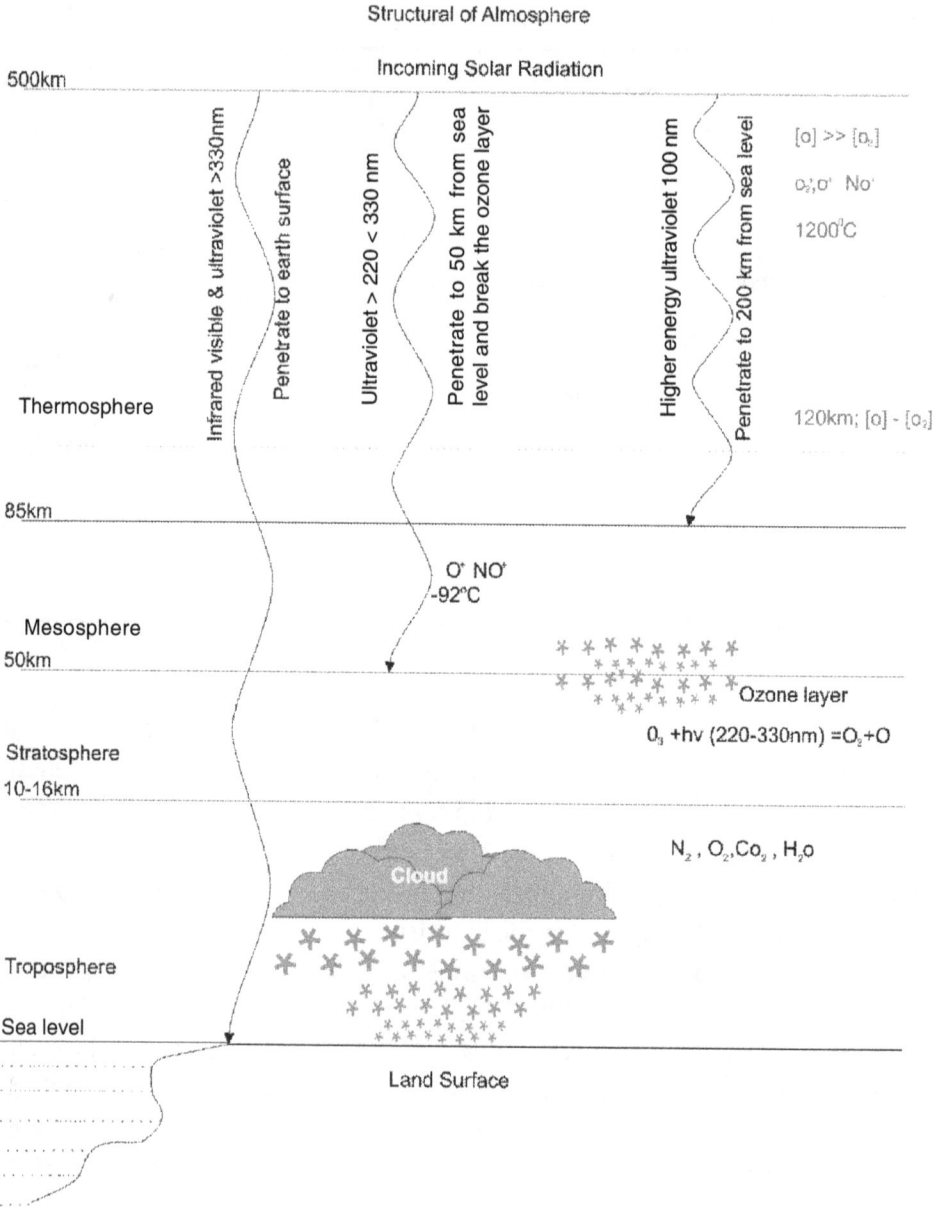

Figure 12.3: Structure of Atmosphere Showing Different Spheres with Traveling of different Radiations and Temperature and Gases.

Troposphere

It is the weather occurring layer extends from Sea level to 10-16 km is comprised of clouds and gases like N_2, O_2, CO_2 and H_2O, where infra-red, visible and ultraviolet radiation of greater than 330 nm wavelength penetrate and reaches to earth surface. This is the layer in which changes in weather conditions at a given time and space occurs due to changes in temperature, pressure precipitation wind types and clouds.

Stratosphere

Known as Aircrafts fly stable layer occurs from 10-16 to 50 km strip is the stratosphere of which above part the ozone layer is present. This ozone layer is in the process of thinning and repairing.

$$O_3 + hv \text{ (220-330 nm)} = O_2 + O$$

Mesosphere

Meteors or rock fragments burn up layer extends from 50 to 65 km strip is the mesosphere from where the ultraviolet radiation of wavelength between 220-330, penetrate to 50 km from the Sea level and breaks the ozone layer, for that the adequate concentration of oxygen is required for its repairing. In absence of adequate O_2 due to green house gases leads to thinning or even hole in ozone layer. It is the coolest sphere with temperature around $-92°C$ in presence of O_2^+ and NO^+ gases.

Thermosphere

Layer with aurora of space shuttle orbits from 85 or 120 km to 500 km, a broadest strip is the thermosphere up to which higher energy ultraviolet radiation of less than 100 nm, penetrates down to 200 km from the Sea level. It is the hottest strip with temperature round 1200°C, where concentration of [O] is much higher than [O_2] as well as O^+ and NO^+ are also present above which upper limit is known as *Exosphere.*

Presence of green house gases, resulting in green house effects that is *" Mechanism that explains atmospheric heating caused by increasing in green house gases is believed to act like the glass in the glass house, permitting visible light to penetrate but impeding the escape of infrared radiation, or heat."*

Arrhenius who received Nobel Prize in 1903 for his work in electrical conductivity had also given his theory on green house effect but unfortunately no attention was given during his life time though, he applauded the possibility of global warming.

Finally, **Revelle** worked on CO_2 and climate modifications, oceanographic exploration and biological effects of radiation in the marine environment and population growth and global food supply received the National Medal of Science in 1991 is said to be the grandfather of "Green House Effect."

It happens due to thinning of ozone layer *"Thin layer of ozone molecules (inorganic molecules) in the stratosphere, which absorbs ultraviolet light and convert it to infrared radiation, screens out 99 per cent of the ultraviolet light effectively.*

12.4 Depletion in Stratospheric Ozone Layer

UV- radiation splitting the Ozone as,

$O_3 + UV \longrightarrow O + O_2$ this is further repaired if, oxygen is present,

$O + O_2 \longrightarrow O_3 + heat$

12.5 Activities Depleting the Ozone Layers

In addition to greenhouse gases, the major reasons of depletion are the long life span gases, due to either their uses in deep freezes, air conditioners and spray or travelling of Jet through stratosphere.

1. Use of chlorofluorocarbons, a very active chemical compound: Use of this chemical compound in 1951 in US, known as **Freon-11** which served as a propellant along with another similar compound **Freon-12,** commonly referred to as **CFCs.**

Commonly Used Freon

Genetic Name	Use	Chemical Name	Chemical Formula
Freon-11	Spray as propellant	Trichloromono-fluoromethane	$\begin{array}{c} Cl \\ | \\ Cl{-}C{-}Cl \\ | \\ F \end{array}$
Freon-12	Coolant in refrigerators, freezers and air-conditioners	Dichlorodifluoro-methane	$\begin{array}{c} Cl \\ | \\ F{-}C{-}F \\ | \\ Cl \end{array}$

Even a single chlorine free radical from a CFC molecule can destroy 1000,000 molecules of ozone. This chlorine atom is formed due to striking of UV-radiation to CFC molecules.

2. Flying of High- Altitudes Jets: Jets engines release nitric oxide (NO) which reacts with ozone to form NO_2 and O_2.

    $$NO + O_3 \longrightarrow NO_2 + O_2$$

    Depletion of Ozone layer was first noticed by **Rowland and Molina** in 1974 who received the **Nobel Prize.**

12.6 Green House Effect

The green house effect is the best described in terms of the annual global average radioactive energy budget of Earth- atmosphere system. The Earth- atmosphere system radiates about 236 Wm^{-2} (long-wave radiation) to space. This amount balances the incoming shortwave radiation from the sun. At a temperature 15^0C, the Earth's surface radiates about 390 W^{-2} of energy. The radiation of the long –wave radiation to space as a result of the intervening atmosphere is referred to as green house effect.

The GHGs double roles, it devours the ozone molecules, that results in thinning and hole in ozone layer and then allow the UV-radiation to pass and travel to the earth surface on one way while creates obstruction in giving free-passage to refraction which results in further heating of the earth surface and finally increases in temperature of troposphere.

12.7 Climate Change and Agriculture

Projected changes in the global climate including rise in ambient temperature, unpredicted rainfall pattern, high atmospheric CO_2 concentration, high intensity short wavelength solar radiation results in low productivity of the crops on one way and incidences of pests and diseases on other.

Climate change can involve incoming of UV-B radiation, alternation in temperature, precipitation and rise in the Sea level due to melting of glaciers.

Though, it may affect C_3, C_4 and CAM plants differently but overall downward trends in productivity of major cereals in tropical countries has generated concerns about the impact of climate changes on crop production system and food security. Even the productivity of major crops (wheat and soybean) of USA will not escape.

Chapter 13
Photosynthetic Stress

Life on this living planet, the Earth is dependent on the processes of photosynthesis performed by the plants, the one and only source of different forms of energy. Between sun and human, it is the plants, which act as an industry to manufacture food for our survival.

"Photosynthesis [GK. *Photos*, light + *syn*, together + *titbenai*, to place]: The conversion of light energy to chemical energy; the production of carbohydrates from carbon dioxide and water by using light energy in the presence of chlorophyll."

It includes processes other than photorespiration and CO_2 fixation. Nitrate and nitrite reduction, sulfate reduction and amino acid, lipid and other biosynthetic activities are all the important light driven metabolic processes.

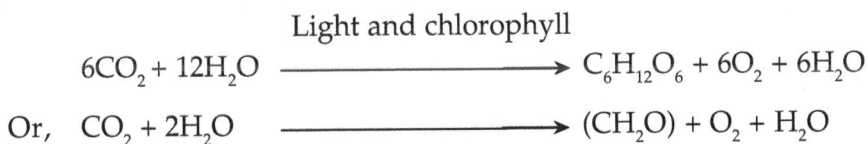

$$6CO_2 + 12H_2O \xrightarrow{\text{Light and chlorophyll}} C_6H_{12}O_6 + 6O_2 + 6H_2O$$

$$\text{Or,} \quad CO_2 + 2H_2O \xrightarrow{\hspace{3cm}} (CH_2O) + O_2 + H_2O$$

Therefore, photosynthesis is the synthesis of solar energy into carbohydrates from carbon dioxide and water with release of oxygen. The green pigment (chlorophyll) of the plant uses solar energy to oxide water and releases oxygen, and to reduce carbon dioxide, thereby formation of large carbon compounds, primarily sugar occurs.

Optimum level of all the requirements of photosynthetic indices is a must for normal functioning of the bio-chemical reactions.

13.1 Response of Photosynthesis to CO_2 Concentration

Since photosynthesis is an oxidation reduction reaction, in which net assimilation is a function of CO_2 concentration at the site of Rubisco (Ribulose, -1,5 biphosphate-carboxylase- oxygenase) in the chloroplast. With rising CO_2, there is no net CO_2 assimilation, until the production of CO_2 in respiration is fully compensated by the fixation of CO_2 in photosynthesis. Thus the CO_2 concentration at which this point is reached is termed as **CO_2- compensation point.** Therefore, optimum concentration of CO_2 is required for its maximum fixation which varies in C_3, C_4 and CAM species differently. In C_3 plants this is largely determined by the kinetic properties of Rubisco, with value of CO_2 compensation point in the 40-50 m mol (CO_2)/mole (air) at 25°C atmospheric pressure.

CO_2 is a substrate for the primary carboxylase of atotrophic organisms (Rubisco). This enzyme is the entry point for inorganic carbon in to the

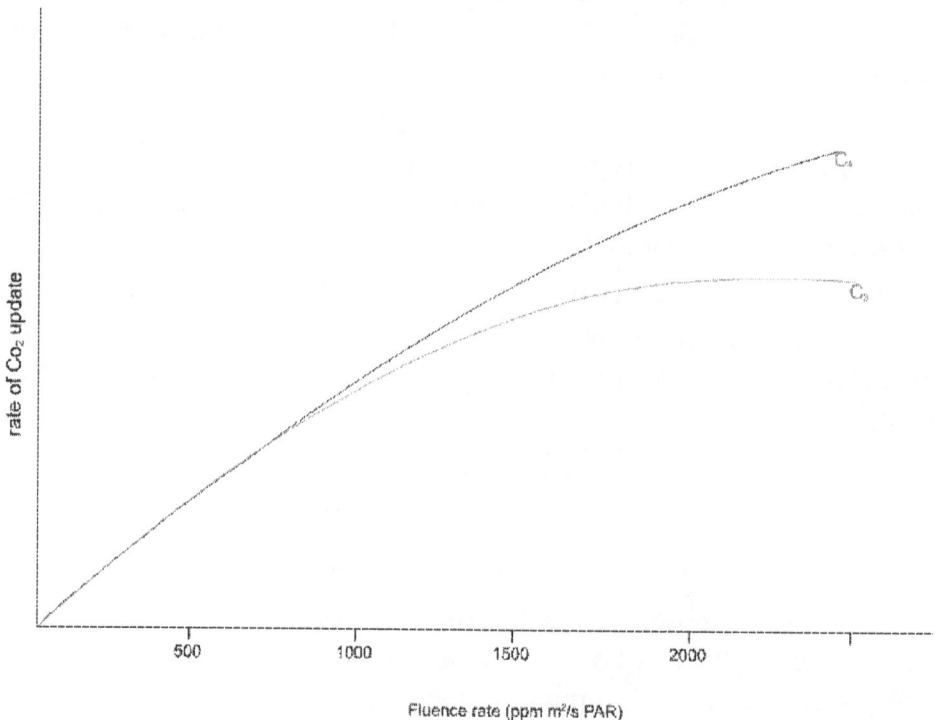

Figure 13.1: Relationship of Light Wave and CO_2 Fixation in C_3 and C_4 Plants.

photosynthetic carbon reduction (PCR) cycle and the organic biosphere. C_3 plants which constitute 95 per cent of the plant species, Rubisco is a major component regulating CO_2 assimilation. The affinity of Rubisco for CO_2 varies from species to species. Terrestrial C_3 and C_4 plants have Km values for about 10 and 30mM CO_2, respectively. Microalgae and submerged plants have higher values (30-70mM) and Cynobacteria have the highest values (80-300 mM). Therefore, present day CO_2 concentration is still not enough to saturate the Rubisco.

13.2 Response of Photosynthesis to Light

There are four classes of photoreceptors n plants. The *photochromes* absorb red® and far-red (FR) light (Ca.660 and 735 ηm, respectively) and have a role virtually in every stage of development from seed germination to seed formation. *Cytochromes* and *phototropines* detect both blue (400-450 nm) and UV-A (320-400 ηm) radiations. The photochromes play major roles during the seedling development, flowering and resetting the biological clock while phototropines mediate phototropic responses or differential growth in a light gradient. The fourth classes of photoreceptors that mediate responses to low level of UV-B (280-320 nm) light have not been recorded.

Acidification in plants needs irradiant energy within the light wavelength of photoactive radiation or visible spectrum in the range of 400-700 ηm where tropical and arid plants assimilate nearer to shorter wavelength 400 ηm of high energy value while temperate nearer to longer wavelength 700 ηm of low energy values (Table 12.1). Low light intensities pose stresses on plants because lower irradiant limits photosynthesis and thus plant growth due to low net carbon gain. High light intensities may also be a stress for plants due to damaging of photosynthetic apparatus and even death of plants. Due to certain genotypic characteristic, the plants are known as **shade plants** and **sun plants**.

Photo damage caused by the excess irradiance is the direct effect of light on the O_2 – evolving complex and thus photo damage to PSII occurs resulting in loss of assimilation.

13.3 Effects of Temperature on Photosynthesis

13.3.1 C3 Plants

At high temperature the oxygenating reaction of Rubisco increases more than the carboxylating one so that photorespiration becomes more important due to decline in solubility of CO_2. The effect of high temperature on C_3 plants is due to its effects on kinetic properties of Rubisco. It affects solubility, affinity and mesophyll conductance and finally increase in photorespiration and reduced yield.

Low temperature causes feedback inhibition due to limitation in sucrose metabolism and/or phloem loading. Chilling injury at low temperature (10-20°C) and frost injury at 0°C in tropical species are quite common. According to another study, at temperature above 15°C accompanied with low CO_2, the assimilation rate is reduced in C_3 plants but at high temperature it continued to rise even above 30°C while, low light intensity and high temperature is detrimental for assimilation. However, temperature up to 30°C associated with high light intensity results in higher assimilation rate in C_3 species.

13.3.2 Prediction of the Photosynthetic Response of C_3 Plants

Prediction of the photosynthetic response of C_3 plants to varying temperature at a given CO_2 concentration is based on the kinetic properties of Rubisco.

$$\frac{Vo}{Vc} = \frac{[O_2]}{Srel[CO_2]} \qquad \qquad ....(1)$$

where,

Vo: Ratio of oxygenase

Vc: Ratio to caboxylase

Srel: Relative specific factor

If, it is assumed that under light saturation, Rubisco is saturating and photosynthesis is limited by Rubisco then assimilation (A)

$$A = \frac{Vc[CO_2](1-0.5\,Vo/Vc)}{[CO_2]+Kc[1+(O_2)]/Ka} - Rd \qquad \qquad ....(2)$$

where,

Vc: Velocity of carboxylation at substrate solution

Kc and Ko: Michalis constants for CO_2 and O_2

Rd: rate of respiration in the dark, is used an estimate for mitochondrial respiration with a Q10 of Calvin 2.

With increasing temperature the v_o/v_c ratio increases due to a decrease in Srel and increase in the $[O_2]/[CO_2]$ ratio

With increasing temperature V_C and K_C increase simultaneously.

Therefore, C_3 plants have a flat and broad temperature response curve for photosynthesis.

At a given temperature, v_o/v_c can be computed using inputs of both Srel, $[CO_2]$ and $[O_2]$.

$$\text{Or, } v_o/v_c = 2 \, I^x/[CO_2] \qquad \qquad ....(3)$$

where,

I^x (in Pa) is the compensation point for photorespiration in absence of dark respiration, which is temperature dependent.

$$I^x = P[42.7 + 1.68 \, (T - 25) + 0.012 \, (T \, 25)^2] \qquad ....(4)$$

where,

T: Temperature (°C)

P: Atmospheric pressure (MPa)

This model predicts a broad temperature response to photosynthesis at current ambient CO_2 concentration and increase in temperature with higher CO_2 concentration will change the kinetic properties of Rubisco. It shows how interaction between changes in CO_2 levels and global warming could affect the photosynthesis in C_3 plants.

In C_3 plants, temperature of 20-30°C under low to moderate light intensity, photosynthesis is dependent on Rubisco which is also approximately coincides with the electron transport for leaf photosynthesis.

13.3.3 C_4 Plants

These plants normally operate at saturating concentrations of CO_2 and therefore, Rubisco is more sensitive to temperature than in C_3 plants.

Therefore, photosynthesis shows the strong temperature dependence under high light with a Q10 of around 2 but it never occurs at low temperature. There is a significant rise in photosynthesis in C_4 plants with increasing temperature under high light intensity.

C_4 plants are sensitive to low temperature and sudden drop in photosynthesis occurs below 12°C. Stomatal conductance in maize growing under chilling temperature is not reduced but reduction in mesophyll conductance occurs.

C_4 grasses are very sensitive to low temperature (20°C) and rise in temperature (30°C) increases the photosynthetic rate while chilling of maize plants under high light results in irreversible loss in CO_2 fixation due to increase in ATP levels in leaves, which damages the cells.

High temperature stress in C_4 plants may cause inactivation of certain enzymes of the Benson – Calvin Cycle in leaves. Rubisco and PEPC have a high thermal stability at 50°C in sorghum and peanut as compared to maize and even soybean (C_3).

As the thermal limits are exceeded above 40-50°C, C_4 plants are susceptible to photosynthesis and a drop in CO_2 fixation. Anyway, C_4 plants are more tolerant to high temperature and thermal break points than C_3 plants.

13.3.4 Predicting the Temperature Dependence of C_4 Plants

Photosynthesis in C_4 plants is a function of Rubisco close to CO_2 saturation. At 30°C with moderate to high levels of radiant energy, the CO_2 concentration in bundle sheath is estimated to be 25-70ppm. This results in a v_c/v_o ratio for Rubisco caboxylation/oxygenation of calcium from 8:1 to 20:1 and a rate of CO_2 production by photorespiration which is only 2 to 6 percent of Vc. Increasing temperature from 15 to 40°C has no effect on quantum yield of CO_2 fixation in C_4 plants and net rate of CO_2 uptake.

If Rubisco functions under saturating levels of CO_2 over a range of temperatures, then the temperature response of C_4 plants may be dependent on Kc of Rubisco and unaffected by Rubisco oxygenase. Under saturating levels of Ribulose biphosphate (RuBP) at high levels of light energy in absence of photorespiration, the eq. (2) used for C_3 photosynthesis may be;

$$A = \frac{Vo[CO_2]}{[CO_2]+Kc} - Rd \qquad \qquad ....(5)$$

If the CO_2 concentration in the bundle sheath cells is highly related to Kc, then the value of Kc will have limited effect on the rate of CO_2 fixation. Under high RuBP and CO_2, photosynthesis rates in C_4 plants would be controlled by temperature dependence of Rubisco, which has a strong temperature dependent under high light intensity.

In summary, photorespiration is relatively low in C_4 plants, it may be temperature dependent of photosynthesis under water stress conditions when the intracellular levels of CO_2 are a limiting factor.

13.4 Effect of Temperature on CAM Photosynthesis

High optima for day time photosynthesis and low optima for night time acid accumulation are the two distinct optima occur in CAM plants. Several facultative CAM plants functions in a C_3 versus CAM mode, high day temperature/low night temperature favours the CAM mode. In these plants, PEPC increases by three fold in the Km (PEP) between 15°C and 30°C and reduces the Ki (malate) between 25°C and 35°C, the activation of enzymes favours at low temperature. Higher temperature during night increases CO_2 evolution and accordingly reduces net assimilation.

13.4.1 Effect of Water and Salinity Stresses on Photosynthesis

Globally agriculture and industry are the major users of water (92 per cent) most water comes from surface water supplies -rivers, streams and lakes. Virtually almost all the nations experiences shortage of water which is predicted to become worse by 2025. The shortage is estimated to be 2.6-3.1 billion as compared to 434 million of present day. In several less developed countries together with in arid to semi-arid countries do not have facility to clean or even drinking water.

Unfortunately, rainfall is not evenly distributed across the world. As such, tropical rainforests receive 250 cm or more rain while only 25cm in desert each year. According to world watch Institute between1950-2050, there will be 74 per cent fall in available water for each person.

If the rainfall is less than 70 per cent during 21 days or longer, it is termed as drought conditions. Drought also causes loss in ground water resulting unavailability in drinking water and loss in forest area. Restoration of watersheds and wetlands for mitigating water shortage in arid and semi-arid regions may prove to be a life line.

Checking population growth on one hand and conserving water in watershed areas in lakes and tanks are the need of the hour on the other. While protecting the environment restoration of vegetation in watersheds, replanting forest and grasslands to reduce sedimentation in streams and reservoirs leads to harvest water efficiently.

13.4.2 Effect of Plants Nutrients on Photosynthesis

Since all the 16 plus nutrient elements are essential for the growth and development of the plants hence, it is but natural that they may have their roles in photosynthesis directly or indirectly. The Hill and Rubisco activities are closely related to concentration of N in leaves as well as its role in PSI and PSII and other pigment-protein complexes (Cytochrome f and coupling factor). Therefore, photosynthesis is very strongly affected by N availability as more than half of the leaf-N and larger portion of the rest is directly and indirectly plays role in photosynthesis. It is more pronounced in C_4 plants than C_3 and it also varies within the C_3. The photosynthesis rate to per unit of N or Photosynthetic N-use efficiency (PNUE) at the growth irradiance may be highest in the leaves with low N-concentrations due to higher degree of utilization of the photosynthetic apparatus. The CO_2 fixation under low N is down regulated due to decreases in Rubisco, chlorophyll and stomatal conductance.

N-concentration to per unit leaf area of the leaves facing the sun-light is usually found higher to those under the shade down the canopy. Therefore, upper younger leaves with higher-N content may have higher CO_2 assimilation. On the other, if the leaf-N content is too high growing under low light intensity, only a fraction of the photosynthetic capacity is utilized. Deficiencies of both N and water have a marked reduction in carbon fixation.

Soil with low-P, photosynthesis may show little correlation with tissue-N, but a strong correlation with tissue P-concentration. Plants grown on low-P usually show low photosynthetic rate due to feedback inhibition caused by low P-concentration in the cytosol or low concentration of Rubisco and other enzymes.

In soybean, failure in tracking of solar energy under P-deficient soils occurs while stomatal conductance in rice and maize in absence of adequate N and P concentration has been established. In white pine (*Pinus strobes* and *Pinus taeda* L.), P and N nutrition is correlated with photosynthesis. The magnitude of photosynthetic response to high CO_2 concentration is fully dependent on the soil fertility and nutrient supplying power to plants. It is also true that plants growing under nutrients stress conditions will response much less to elevated CO_2 than those well nourished.

In *Gossypium hirsutum*, the net CO_2 fixation is reduced at low concentrations of N, P and K due to thinner leaves formation followed by lower concentration of chlorophyll. Even deficiency of Mn affects PSII reaction centre and leaves with low chlorophyll content.

13.4.3 Photosynthetic Classification of Plants

On the basis of stable primary products of photo synthesis the plant species have been broadly classified in to three types: C_3, C_4 and CAM. If, the stable primary product of photosynthesis is the 3–carbon intermediate or compound, phosphoglyceric acid (PGA), the plant is known as C_3 species while if, the stable primary product of photosynthesis is the 4-carbon intermediate or compounds, aspirate and malate or oxaloacetic acid (OAA), the plant is known as C_4 species.

The third type plants species is the intermediary of both C_3 and C_4 plants is termed as Crassulacean acid metabolism (CAM). It is the photosynthetic pathway in which the first step of CO_2 assimilation is the carboxylation of phosphoenolpyruvate (PEP) by PEP carboxylase; the first stable product is oxaloacetic acid (OAA) - a four-carbon intermediate; in contrast to C_4 photosynthesis, the CO_2 assimilation occurs predominantly during the night with open stomata.

High activity of PEP- carboxylase enables C_4 plants to reduce the stomatal aperture and thereby conserve water while fixing CO_2 at rates equal to or greater than those of C_3 plants. Suppression of photorespiration resulting from the concentration of CO_2 in the crantz cells also increases the photosynthetic efficiency. These characteristics make C_4 species to photosynthesize more efficiently at high temperature and sustain in dry-hot conditions. Therefore, water requirement (Table 13.1) of C_3 plants to per unit of CO_2 fixation is higher than C_4 and even minimum of CAM species (cactus).

Table 13.1: Loss of Water (g) for/g of CO_2 Fixation.

Plants	Water Loss g/g of CO_2 Fixation
CAM	50 – 100
C_4	250- 300
C_3	400 – 500

It is very clear that water requirements of CAM: C_4: C_3 plants for each unit of CO_2 fixation are almost in the ratios of 1: 4: 6.

At higher temperature the quantum yield remains constant in C_4 plants due to low rate of photorespiration but it decreases in C_3 plants due to higher energy demand for per net CO_2 fixation.

13.5 Climate Change and Photosynthesis

As discussed, there sould be the optimum climatic conditions; CO_2 concentration, temperature, radiation and other essentiality for normal photosynthetic processes in plants. Since, there is hardly any possibility for decreases in all these climatic parameters hence, rises in ambient temperature, CO_2 concentration, high intensity radiant energy, presence of ozone in the lower sphere and several other pollutants may affect the productivity of the crop industry.

13.5.1 Effect of High Temperature

Crop response to temperature depends on the specific optimum temperature for photosynthesis, growth and yield. If, the temperature is below optimum for photosynthesis, a slight increase in temperature may result an increase in plant growth and development, but if, temperature is close to maximum, a minor increase in temperature can affect the crop negatively and with decrease in yield.

1. Temperature above 30°C is detrimental for bean, linseed, tomato, maize, wheat and *Brassica* species.
2. Even a brief period of higher temperature at reproductive stages of groundnut and wheat results in low yields due to sterilization of pollen. It also reduces protein content in all cereals.

3. High temperature increases abscisic acid (ABA) but decreases indole-3-acetic acid (IAA) and ethylene evolution resulting in low crop productivity since IAA is a growth promoting hormones.

13.5.2 Temperature and Stomatal Conductance

Higher temperature causes stress which is a failure cause in supply of one component part of regulatory network due to overloading. Higher temperature results in a high vapour pressure deficiency (VPD) in stomata by an increase in leaf temperature under low humidity in air, cause stomatal closure, and thus limit the supply of CO_2 for photosynthesis.

If low soil moisture content is associated with high temperature to stomatal conductance, can decrease and slow down the photosynthetic process. Water deficit combined with high temperature is more accountable in decreasing the photosynthesis than water stress at low temperature.

13.5.3 Temperature and Metabolism

The changes in metabolic process in plants may be of two types.

1. Changes due to temporary variation in temperature
2. Changes due to permanent change in temperature

Temporary changes in temperature may result short term changes in leaf temperature, results changes in proton flux density while permanent changes in temperature may result changes in gene expression.

Changes in temperature also affect enzymes activities. It also changes the intracellular pH that affects calcium nutrition.

13.5.4 High Temperature and Photosynthesis in C_3 Plants

High temperature increases the oxygenating reaction of Rubisco more than the carboxylation so that the photorespiration becomes proportionally more important due to decline in solubility and affinity for CO_2 and enhancement in release of O_2. The electron transport may also decline at elevated temperature. The combined effects of temperature on affinity, solubility and mesophyll conductance cause a proportional increase in photorespiration and thus finally net decline in photosynthesis.

Species adapted to hot environments usually show temperature optima for photosynthesis that are very close to temperature at which enzymes are inactivated. Therefore, at high temperature both Rubisco activity as well as membrane-bound processes of electron transport may limit the photosynthesis process and less crop yield.

High temperature can cause both reversible and irreversible effects on photosynthesis in both C_3 and C_4 plants. Inhibition in photosynthesis in peas at 20°C and in wheat and cotton at high temperatures was recorded due to increase in photorespiration and inaction of Rubisco. The high temperature limits the Rubisco activity which is reversible between 40-50°C due to destruction of electron transfer to reaction centres from water pigments and thus restriction or rather complete inhibition in photosynthesis occurs.

In C_3 plants (*Brassica* species), high temperature inhibits chloroplast biogenesis by causing deficiency in chloroplast ribosomes but not in C_4 plants (*Zea mays*).

High temperature can impair the phloem translocation and sink development in several C_3 species. In wheat and barley endosperm and in potato, tubers reduce the synthesis of starch. In barley, temperature above

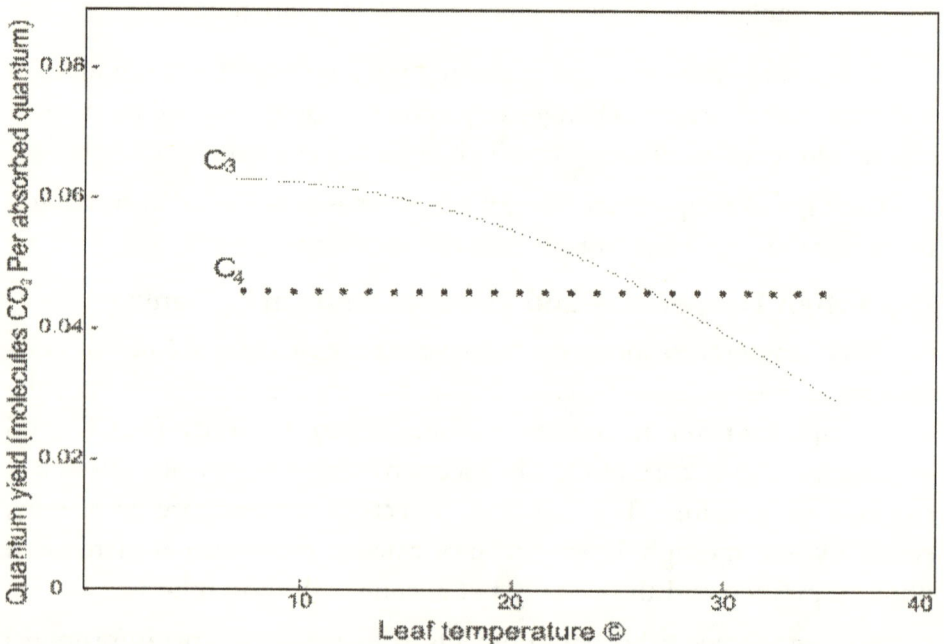

Figure 13.2: Quantum Yield of Individual Leaves as a Function of Temperature for C_3 and C_4 Species.

30°C reduces synthesis of sucrose. Impairment in starch synthesis in wheat at 35°C may be related to reduction in synthesis of soluble starch. Potato tubers have temperature optima of 21.5°C for change of sucrose to starch.

13.5.5 High Temperature and Photosynthesis in C_3 and C_4 Plants

The CO_2-compensation point of C_4 plants is only 0-5m mole/mole of CO_2 while it is 40-50m mole/mole of CO_2 in C_3 plants. Absence of photorespiration in C_4 plants is one of the positive characteristics for higher yield. At 380ppm of air CO_2 concentration, the internal concentration in the mesophyll of C_4 plants is only 100ppm as compared to 250ppm in C_3 plants.

The quantum yield at temperature 30°C and above for C_4 plant is higher to C_3 plants since it is independent of O_2 concentration in the former. At relatively higher temperatures, the quantum yield of photosynthesis is higher for C_4 which is not affected by temperature but quantum yield of C_3 plants declines with increasing temperature due to increase in oxygenating activity of Rubisco (Figure 13.2).

At 21 per cent O_2 and 0.035 per cent CO_2, the quantum yield is higher for C_4 plants than C_3 plants at higher temperature due to photorespiration in the later but at low O_2 concentration and same CO_2 concentration (0.035 per cent) the quantum yield is higher in C_3 species.

The differences in biological and physiological activities also guide the abundance of C_4 monocots (grasses) in warm-season rainfall areas and C_4 dicot in arid and saline soils conditions while C_3 species are more active in cool and moist conditions.

C_4 photosynthesis originated in arid regions of low latitude, where high temperature combined with drought and/or salinity with increased fire frequency, promoted the spread of C_4 plants.

Low altitudes in tropical area, tropical and temperature lowland grasslands with sufficient rain are dominated by C_4 species but at higher altitude of the same area, C_3 species are dominant *viz.*; S- Africa and temperate region of Argentina.

Higher CO_2 concentration and higher temperature are favourable for C_4 species due to absence of photorespiration and therefore, abundance of this species occurs under warm conditions.

In another study, the rate of photosynthesis in C_4 species may be 2 to 3 times higher at 30-40°C than C_3 plants. C_4 plants are better equipped to sustain drought and to maintain active photosynthesis under water stress due to their stomatal closure but under the same situation, carbon fixation in C_3 plants is reduced consequent. Under optimal conditions, C_4 species can fix CO_2 from 2-3 times higher to C_3 species.

13.5.6 Effect of Extreme Temperature

A sudden rise in temperature close to lethal, induces the formation of m RNAs coding for heat- stock proteins which increases the plant tolerance capacity against high temperature. Heat stock protein may be involved in the protection of the photosynthetic apparatus and prevent photo-oxidation and resists milder degree of heat stress. Some of the plants leaves (*Quercus alba*) can increase the above air temperature to 14°C and drops the leaf temperature by 8°C within minutes.

13.5.7 Effect of Low Temperature on Photosynthesis

Proton absorption is not affected by low temperature but the rate of e^- - transport and biochemical processes are reduced. Sucrose metabolism and/or phloem loading may be limiting for photosynthesis, causing feedback inhibition. At extreme cold, many plant (sub) tropical plants grow poorly at 10-20°C, is known as "chilling injury" and differs from frost damage at 0°C. The following changes may occur.

1. Decreases in membrane fluidity
2. Changes in the activity of membrane-associated enzymes and processes such as the photosynthetic e- transport.
3. Loss in activity of cold sensitive enzymes.

Chilling leads to photoinhibition and photooxidation because the biophysical reactions of photosynthesis (proton capture and transfer of excitation energy) are far less affected by temperature than the biochemical stress, including e- transport and activity of the Calvin cycle. At exposure to low temperature and high light, the conversion of the light-harvesting **violaxanthin** to the energy- quenching **Zeaxanthin** occurs within the minutes.

13.5.8 CAM Plants Adaptation to Desert Life

Since majority of this group belongs to Crassulaceae family of which the first stable product of photosynthetic acid is known as Crassulacean Acid and hence the plants belong to this group is termed as Crassulacean Acid Metabolism. Some 10,000 species from 25-30 families of 10,000 flowering plants belong to this class. Some ferns belong to CAM group which includes *Cactaceae* and *Euphorbiaceae*. It includes the lowest water use efficiency and high humidity tropical plants like pine apple to the highest water use efficiency species and high temperature (50°C) adapted desert plants. Agave and cacti have the highest temperature and thermal tolerance and a number of cacti species can tolerate 50°C as compared to other species tolerant limit of 47°C. Even several cacti can survive for one hour at this temperature due to the following characteristics.

1. It consists inverted stomatal cycle- stomata open during night and are usually closed during day, meaning by CO_2 fixation only occur during night.
2. It accumulates malate at night and subsequently deplete in the day.
3. Nocturnal stomatal opening supports a carboxylation reaction producing C_4 acids which are stored in watery vacuoles.
4. Transpiration ratio of CAM plants is lower (50 to 100) than either C_3 or C_4 plants but CO_2 fixation is 50 per cent of C_3 and 33 per cent of C_4 however, CAM are more drought resistant among all due to retaining and re-assimilation of inspired CO_2.

Though, CO_2 assimilation by CAM is slow but their higher WUE allows photosynthesis to continue in water-stress conditions and complete the reproductive phase.

13.6 Effect of Elevated CO_2 Concentration on Photosynthesis

When plants made the transition to this planet, the atmospheric CO_2 concentration was up to 16-fold higher than today and since the last glacial maximum atmospheric CO_2 concentration has more than doubled to 387 ppm due to industrial revolution in the late 18th century, deforestation, ploughing of prairies, drainage of peat's for increasing human demands but it is still low to even saturate the photosynthetic activity of C_3 plants.

It is increasing @1.5 ppm/year A further doubling of CO_2 concentration, may reduce the O_2 inhibition of Rubisco and halve photorespiration; reduce stomatal conductance and enhance water-use-efficiency; increase the C/N ratio; lower dark respiration, exert growth modulator effects and inadequate species influence the photosynthetic affinity or dissolved inorganic carbon.

According to another study, the stomatal conductance and atmospheric CO_2 concentration are negatively correlated. It also decreases the stomatal density, reduces biological discrimination for ^{13}C of the fossil leaves and some other herbarium specimens may increase their water-use-efficiency under increased CO_2 concentration. A general decline in herbs nutritional concentrations over the last 250 years including C/N ratio and changes in mineral content were also recorded due to rise in CO_2 concentration which has also influenced the soil nutrients status.

13.6.1 Species Differences to CO_2-Enrichment

Plants with higher sink capacity have the highest response to CO_2-enrichment with 30-40 per cent stimulation of biomass but with small sink and stress tolerance species may have the least advantage.

Among C_3 species, the response to CO_2-enrichment is variable. Existence of inherent site-specific and intra-specific differences is reflective of RuBP regeneration and sink capacity. CO_2-enrichment causes growth regulatory effects on photosynthesis cannot be ruled out.

C_4 plants are adapted to hot-dry climate with about 1,500 species belonging to 18 angiosperm family (3 monocots and 15 dicots) since they have mechanism to avoid the impact of photorespiratory CO_2 loss by concentrating CO_2 in the carbon-fixing cells. These plants have a low CO_2 compensation concentration and low transpiration ratio which means these are able to maintain higher rate of photosynthesis at low CO_2 levels.

In C_4 leaves the vascular bundles are very close together and each bundle is surrounded by a tightly fitted layer of cells called the *bundle sheath*. Between the vascular bundles and adjacent to the air spaces of the leaf are the more loosely arranged *mesophyll* cells which distinguishes between mesophyll and bundle sheath photosynthetic cells, known as *Kranz anatomy* (*wreath, in German* is the 'spongy", mesophyll cells), plays a leading role in re-trapping of the to be released CO_2 in the C_4 synthesis.

13.6.2 Effect of Elevated CO_2 and Temperature

In some of the studies under laboratory conditions, though doubling in CO_2 concentration increased the initial growth of the plants but no advantage in final yield was recorded due to increase in photorespiration as the temperature also increased simultaneously. Results of the some limited works done under controlled conditions in the glass-house indicated (Figure 13.3) that;

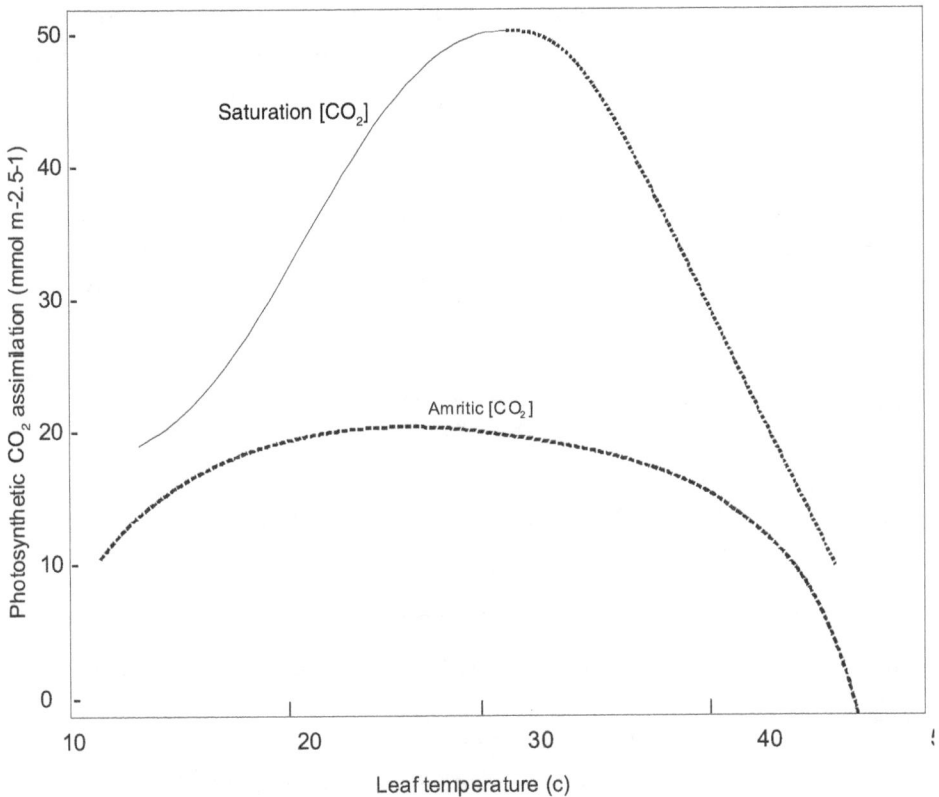

Figure 13.3: Photosynthesis as a Function of Temperature and $[CO_2]$.

CO_2 elevation is also favorable for its fixation but temperature beyond 35°C is injurious

1. CO_2 enrichment and thereby increase in temperature reduced the Rubisco content in rice and soybean

2. Even increased nutrients and water failed to mitigate the adverse effects of enrichment in CO_2 above optimum temperature on the food and fibre productivity

3. Besides elevation in CO_2 concentration and temperature, global warming may change the precipitation patterns, resulting occurrences in unpredicted flood and drought. Therefore, considering only CO_2 and temperature simply can not address the whole issue of global climate that will finally affect crop production in future.

4. Other abiotic and biotic factors including UV-B radiation, ozone concentration, relative humidity, wind velocity and several other pollutants are a big question to answer for future generations.

Therefore, a future research thrust is needed on the crop responses to combined effects of rising CO_2 and temperature (Figure 13.4).

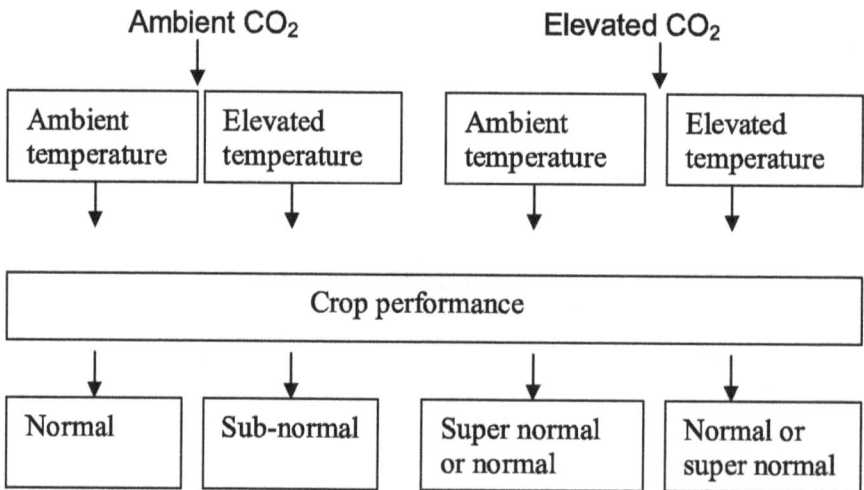

Figure 13.4: Crop Responses to Combined Effects of CO_2 and temperature.

13.7 Effect of Elevated Light on Plants

Too much light inhibits photosynthesis, is known as photoinhibition of photosynthesis and is termed as the light dependent decrease in photosynthetic rate. This occurs under excess irradiance which is needed for photosynthetic evolution of O_2. While under limited light conditions the rate of CO_2 assimilation is also limited and is saturated under adequate light but excess light results in photoinhibition or decrease in net photosynthesis.

Among the two phases (PS I and PS II) of photosynthesis, PS II is more sensitive to photoinhibition than PS I. Photoinhibition that results in

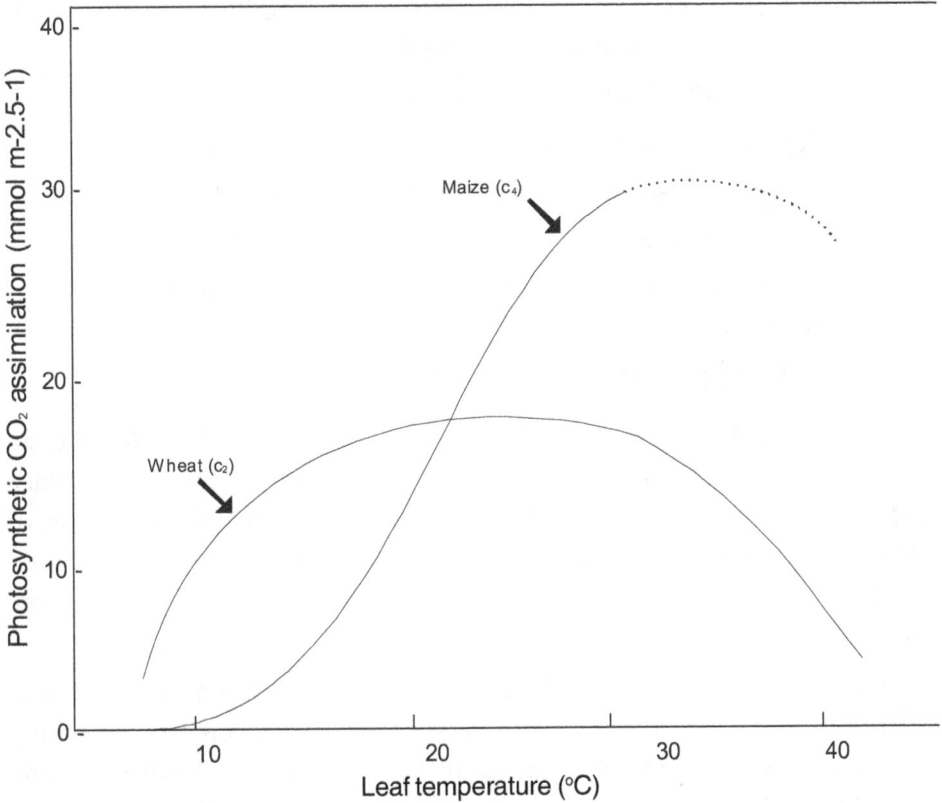

Figure 13.5: Photosynthesis in C_3 and C_4 Plants as a Function of temperature.

decreases in photosynthetic efficiency as well as photosynthetic capacity results in **chronic photoinhibition** which is the result of photo damage to PS II and net-photosynthesis.

Higher leaf temperature of about 30°C and lowest at 15°C is ideal but temperature above 25°C is injurious for yield performance in C_3 plants while higher leaf temperature at 35°C and lowest at 20°C is the optimum range for C_4 plants. is desirable (Figure 13.5).

13.7.1 Effect of UV-B Radiation

UV-B radiation which passes through the stratospheric ozone shield has very detrimental effects on plants. The inhibition of photosynthesis or electron transport under excess light or UV- irradiation may elevate the photosensitization process as well as AOS (Active Oxygen Species) in the following ways.

1. CO_2 fixation in Mehler reaction may be inhibited up to 50 per cent due to production of more H_2O_2
2. Both the PS I and PS II systems may be activated
3. Degradation in D1 and D2 proteins may also occur
4. Loss of Rubisco causes inhibition of CO_2 assimilation
5. Damage to DNA
6. Alternation in transpiration, photosynthesis and respiration potential.
7. Reduced growth, development and morphology

Since 1980, the decrease in stratospheric ozone layer by 3-6 per cent, resulted in 6-14 per cent increase in UV-B radiation at the Earth surface (WMO, 2003). Depletion in ozone layer, resulting in huge emission of CFCs, CH_3Br and NOx by human activities which is further intensified by clouds, aerosols, season, location and different gases including tropospheric ozone itself.

Presence of ozone in the Earth's stratosphere prevent all the UV-C (<280 ηm) and most of the UV-B (280-320 ηm) radiations from reaching the Earth surface. Due to differences in optical density of the atmosphere, the UV-radiations reaching the Earth is the least at sea level in polar regions and the greatest at the high altitude and low latitude. Cloud cover also reduces the solar UV-irradiance to a greater extent.

DNA is very sensitive to UV-radiation which results loss in biological activity. Since the concentration of RNA is much higher than DNA hence the effects on RNA is less than on DNA.

Algae and bacteria are more sensitive to UV-B radiation than the leaves of higher plants due to less shielding of their DNA. Smaller internodes and small curled leaf formation in higher plants are the common features under UV-B radiation (<300 ηm) resulting in reduced photosynthesis and productivity.

13.7.2 Combined Effect of UV-B Radiation and other Factors

The effects of UV-radiation with other requirement of photosynthetical parameters in plants act in different ways. Some of them elevate the injury made by the extreme radiation while some others may eliminate the adverse effects.

1. The sensitivity of plants to UV-B irradiation is much higher at low PPFD (Photosynthetic Proton Flux Density) and UV-A levels. UV-A reduces the damaged caused by UV-B at low PPFD only.

2. UV-B –induced radiation in plant growth is less severe at high CO_2 concentration.

3. Either elevated CO_2 or higher temperature has similar effect in reducing the growth- inhibitor effects caused by UV-B radiation.

4. Elevated CO_2 concentration may reduce the UV-B effect on physiological parameters.

5. UV-A radiation may decompose the UV-B absorbing pigments and this eliminates the effect of UV-B radiation.

6. UV-A radiation is 7.5 times biologically more effective than UV-B radiation.

7. Plants under water-stress conditions are more susceptible to UV-B radiation.

8. Increases in proline content and decrease in stomatal conductance in Arabidopsis plants protect them to sustain under UV-B radiation in desert climate.

13.8 Impact of Ozone on Crop Production and Quality

Ozone, the tri-atomic allotropic form of oxygen, is a colourless gas with a slightly sweet in taste and watermelon like odor. Stratosphere (16-50 km above ground level) ozone protects the Earth's surface from solar UV-radiation while troposphere (0-16 km above the Earth's surface) ozone is injurious after CO_2 and CH_4 gases.

Crop potential yield depends on (a) defining factors (CO_2, radiation, temperature and crop traits/species), (b) limiting factors (water and nutrients) and (c) reducing factors (pests, pathogens, weeds and pollutants).

Increasing ozone concentration at ambient CO_2 results decline in yields of many species and this negative effect is counter acted by CO_2-enriched atmosphere, due to decrease in stomata opening and ozone flux or increase in the activity of anti-oxidant enzymes.

Presence of O_3 at the line of stratosphere and mesosphere as ozone shield to filter the UV-B and UV-C radiations is beneficial but presence of

O_3 in the troposphere is injurious since it acts as a secondary pollutant as a product of the reactions of primary pollutants (NO_2, SO_2, CO and RH) with sunlight (hv) and thus formation of O_3 acts as a protein modifier.

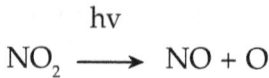

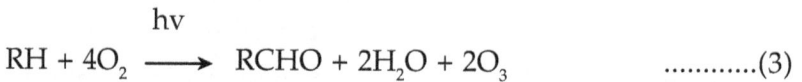

$$NO_2 \xrightarrow{hv} NO + O$$

$$O + O_2 \longrightarrow O_3 \qquad\qquad\qquad ....(1)$$

$$CO + 2O_2 \xrightarrow{hv} CO_2 + O_3 \qquad\qquad ....(2)$$

$$RH + 4O_2 \xrightarrow{hv} RCHO + 2H_2O + 2O_3 \qquad ....(3)$$

Presence of high level of O_3 and SO_2 in the lower atmosphere is injurious since these enter into the leaf stomata and damage the photosynthetic cells directly. Higher O_3 concentration decreases leaf photosynthesis and leaf area and thus finally the rate of photosynthesis while higher concentration of SO_2 causes leaf injury (necrosis and abscission of leaves).

Effects of elevated $[O_3]$ on different crop species resulted 20 per cent reduction in leaf-level photosynthesis at 70 ppb O_3 compared to carbon-filtered air. In barley crop exposed to 180 ppb O_3 reduced photosynthesis by 17 per cent to that grown at 10 ppb O_3 due to adverse effects on stomatal conductance. The effect of high O_3 level on photosynthesis is linked to decrease in amount of Rubisco in peas, radish, potato, clovers and oat due to break down of photosynthetic pigments.

Chapter 14

Respiratory Stress

Intake of oxygen and release of carbon dioxide is termed as respiration and the relationship of CO_2 and O_2 is known as respiratory quotient (RQ). In biological term, respiration is the activity of glycolysis, the oxidative pentose phosphate pathway (OPPP), the tricarboxylic acid (TCA) cycle and mitochondrial electron transport and related oxidative phosphorylation. These biochemical reactions and pathways releases CO_2 at early stage of OPPP and TCA cycle and consume O_2 as terminal steps in mitochondrial electron transport.

Majority of the nonphotosynthetic, photosynthetic and nonphotorespiratory CO_2 and O_2 exchanges in crop plants are related to biochemical reactions of respiration as:

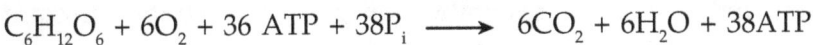

$$C_6H_{12}O_6 + 6O_2 + 36\ ATP + 38P_i \longrightarrow 6CO_2 + 6H_2O + 38ATP$$

Chemically, the plant respiration can be expressed as the oxidation of the 12-carbon molecule sucrose and the reduction of 12, oxygen molecules.

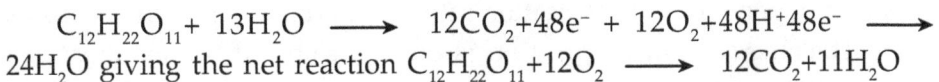

$$C_{12}H_{22}O_{11} + 13H_2O \longrightarrow 12CO_2 + 48e^- + 12O_2 + 48H^+ 48e^- \longrightarrow$$
$24H_2O$ giving the net reaction $C_{12}H_{22}O_{11} + 12O_2 \longrightarrow 12CO_2 + 11H_2O$

Which is a reversal of the photosynthesis process, in which sucrose completely oxidized CO_2 while O serves as the electron acceptor and reduced to water.

In presence of air (O_2), cells consume O_2 and produce CO_2 and H_2O and in absence of air cells produce lactic acid or ethanol. Plants respire roughly half of the daily photosynthetic yield.

Factors Affecting Respiration

a. Plant factors

  1. Species

  2. Growth habit

  3. Type and age of the organs

b. Climatic variables

  1. External oxygen concentration

  2. Temperature

  3. Nutrient and

  4. Water supply

Only green tissues play part in photosynthesis but in respiration all are active for 24 hours. In Herbaceous species 30-60 per cent of daily gain in photosynthetic carbon is lost to respiration which tended to decrease as the age is advanced. According to another study, young plants lose about $1/3^{rd}$ of their daily photosynthetic as respiration this loss can double in older plants.

In tropical rice growing areas, elevated night temperatures, the photosynthetic loss goes from 70-80 per cent due to high dark respiration.

14.1 Effect of Climatic Factors on Respiration

Effect of Oxygen on Respiration

The equation taking in respiration, oxygen is a substrate in this process which can affect plant respiration directly. However, an aqueous solution at 25⁰C, is about 250mM and Km value for O_2 in the reaction catalyzed by cytochrome C oxidase is less than 1mM, therefore, there should not be an apparent dependence of respiration rate on external O_2 concentrations as such. The decrease in respiration is only possible if, atmospheric O_2 concentration goes down to 5 per cent.

Hydrophillic species growing in low O_2 containing water should be vigoursouly aerated to keep O_2 levels high in the vicinity of the roots.

Herbacious species such as rice and Para grass (*Brachiaria mutica*) rely on a network of intercellular air space (aerenchyma) running from the leaves to the roots to facilitate continuous pathway for the movement of O_2 to the flooded roots.

Some *pneumatophores* (*Avicennia* and *Rhizophora*) grow in mangrove swamps have gaseous pathway for O_2 diffusion into the roots.

14.2 Effect of Elevated [CO_2] on Respiration

It appears that release of CO_2 in outer sphere through respiration may be slowed down due to higher concentration of CO_2 in the outer atmosphere and thus the respiration or loss of energy by the plants may be less. In a large number of experiments, a doubling of [CO_2] in the dark, directly inhibited shoot, leaf or entire plant respiration rate from 10 to 50 per cent but at the same time the effect of elevated CO_2 has only a limited or no direct effect on respiration in several crop plants including potato and soybean. Though, there is contradiction on the direct effect of elevated [CO_2] on respiration but it may slow the processes such as biosynthesis, maintenance, transport of nutrients which are responsible for growth and yield.

If, the elevated [CO_2] can reduce the unnecessary component of respiration, will result in a saving of assimilate that may be used for additional economical growth of the plants.

The elevated [CO_2] also has indirect effects on plant performance through changes in photosynthesis, translocation, growth, plant size and/or plant composition due to changes in temperature and availability of water and nutrients.

Since elevated [CO_2] stimulates the rate of photosynthesis, translocation, growth and level of nonstructural carbohydrates, in the same way to respiration hence increases in nonstructural carbohydrates level result in dilution of protein concentration and finally poor crop quality. Therefore, elevated [CO_2] affects respiration and alters;

i. The rates of processes supported by respiration

ii. Stoichiometries between respiration and the processes supported by it

iii. Rates of futile cycling, alternative pathway activity and other forms of wastage.

Elevated [CO_2] resulted in higher plant respiration due to simultaneous increase in temperature which supports the growth and translocation as well as uptake and N-assimilation.

14.3 Effect of Rise in Temperature on Respiration

Short- term increase in temperature is more effective in stimulation of respiration as compared to long-term with usually, a Q10 of 2 to 2.5 may occur due to kinetic effects on the processes using respiratory products. Long term increases in temperature may affect respiration through its effects on growth and maintenance processes.

Long-term Q10 of respiration is generally smaller than the short-term Q10 due to adaption to some degree. However, long-term warming effects may depend on

i. Roles of processes that require respiration as a source of C-skeletons, ATP and/or NAD(P)H.

ii. Specific respiratory costs of these processes and

iii. The amount of ATP produced per unit of substrate respired and extent of any wastage respiration.

14.4 Effect of Elevated Tropospheric Ozone on Respiration

Effects of ozone or its compound on respiratory enzymes or membrane is known as direct effects of O_3 on respiration. Elevated O_3 damages mitochondrial membranes or respiratory enzymes which may slow the respiration rate. Sometimes if, the damage to respiratory enzymes reduced, the effective coupling of electron transport and oxidative phosphorylation and/or reduction of oxygen, the rate of respiration increases with a loss of ADP phosphorylation per unit of CO_2 released.

Extensive damage to leaf cells due to high concentration of O_3 (>200ppm) for more than an hour is known as direct effect of O_3 on respiration.

Higher concentrations of O_3 slow the fixation of CO_2 in leaves and the use of carbohydrates in respiration as well several processes involving respiration. It also directly interferes in translocation of carbohydrates from leaves to other parts of the plant. Higher [O_3] also reduces the root respiration.

Some of the experimental evidences support the indirect activation in respiration rate in leaves due to disruption in leaves membrane structure and possibly other micromolecules but fortunately the cell respond to repairing or replacing the damage structures which needs additional active transport to re-establish metabolic gradients across the repaired membrane.

Chapter 15
Abiotic Stress and Crops

One of the ancient proverb has very rightly highlighted the significance of natural resources:

> *"Earth teaches us patience and love;*
>
> *Air teaches us mobility and liberty;*
>
> *Fire teaches us warmth and courage;*
>
> *Sky teaches us equality and broad mindedness;*
>
> *Water teaches us purity and cleanliness"*

Therefore, proper utilization of earth (soil), air, water and several others natural resources is the only way out for sustaining a normal life on this living planet. Over exploitation of resources may be suicidal for all the creatures on the earth.

Solar energy is the only source of energy which is being transferred by the plant through photosynthesis in presence of proper soil, nutrients, water and several other climatic factors for the use of entire living organisms. The crop response is dependent on different environmental factors, including climatic factors, biological incidences and the biological makeup of the plants (Figure 15.1).

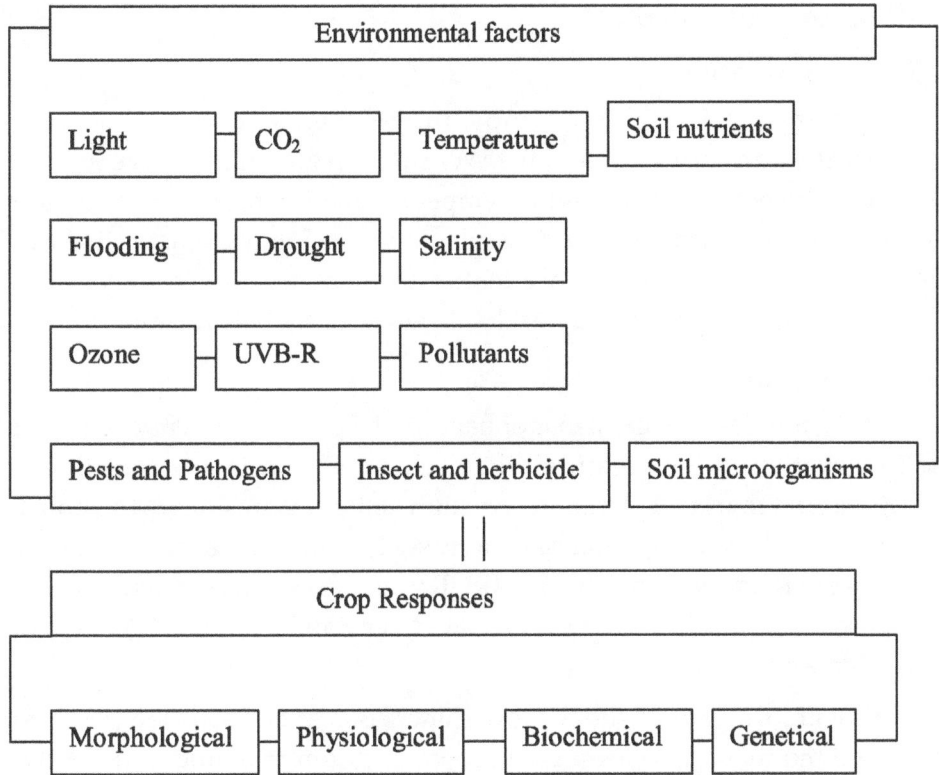

Figure 15.1: Crop Response to Climatic and Environmental Factors.

Crop response is the product of the interactions of plant morphological, physiological, biochemical and genetic makeup with environment and climatic factors (light, CO_2, temperature and nutrient status of the soil). The full potential can be achieved in absence of flood, drought, problematic soils on one hand and ozone, UV-B radiation and other pollutants on the other. Besides these, the incidences of pest and diseases are also accountable.

15.1 Effect of Climate Change on Growth and Development of Plants

Plants meristem temperature is directly responsible for their growth and development rather than air temperature which only influence indirectly, whereas germination and emergence are controlled by soil temperature.

Influence of Elevated CO_2

15.1.1 Germination

Germination is highly a temperature dependent process for both temperate and tropical species. Warm season plants have higher temperature optima compared to temperate one. Tropical species require a temperature optima of 14-18°C while only 3-8°C is needed for majority of the temperate species. Special attention has to be given on the physiology of germination in present day agriculture under global warming.

15.1.2 Growth

Since, the sink metabolism is enhanced, the plant growth rate increases with elevation in temperature while cell expansion is less sensitive to temperature than cell division. At high temperatures, cell division is enhanced and cell elongation rate increases but duration of cell expansion decreases. Relative growth rate (RGR) increases quickly with rise in temperature, almost linearly in sunflower, rape and maize at 28°C to what recorded at 10°C.

Though, the rate of shoot growth increases at elevated temperature, but total bio-mass decreases. Rice grown at different temperature at two CO_2 concentrations, resulted in reduction in bio-mass at 28°C, while higher CO_2 concentration failed to check the reduction due to shortening in growth period. In wheat too, increase in 3.5°C temperature throughout the growing season, 16 per cent reduction in final bio-mass is recorded.

Root growth is also a temperature dependent process. Root diameters are inversely related to root- zone temperature. Higher soils temperature stimulates development of lateral roots. Low soils temperature helps in roots horizontal growth and as such high temperature in vertical growth.

15.1.3 Development

Development progress is measured by the thermal units of degree-days which is expressed as;

$$S = \sum_{day=1}^{day=j} (t - t_0)$$

where,

   S: Sum of degree days (or thermal time)

t_0: The threshold temperature for onset of development (or base temperature) and

t: Mean temperature of the day

Thermal units can be calculated several times during each day (hourly or less).

Shortening of development due to higher temperature in wheat, a 1°C increase in temperature resulted in a 21-day (about 8 per cent) reduction in crop duration.

In rice, flowering is shortened by 10 days when grown at 34/27/31°C (day/night/paddy temperature) in comparison to rice grown at 25/18/21°C. In general faster development results in shorter growth duration. The rate of development is a temperature driven process in day-neutral crops like wheat, maize, cotton, sweet corn, tomato, cucumber, pea and some varieties of rice, sorghum and cowpea.

15.1.4 Effect on Yield Components

Elevated $[CO_2]$ stimulates tillering in cereals. In wheat, 20 per cent increase in grain yield due to 200 per cent increase in tillers has been recorded. In soybean, elevated $[CO_2]$ produced more axillary branches and there by more number of pods and grains. In rice too, CO_2-enrichment recorded 47 per cent increase in yields due to production of more number of panicle per plant.

The variability in seeds size, grown at elevated $[CO_2]$ may be due to variable effects on grain feeling processes. Grain feeling rate increases sometimes while grain feeling duration decreases at elevated $[CO_2]$. Such effects on both processes might be dependent on nutrient supply, temperature and soil moisture. Increases in grain number may increase the demands for nutrients, specially nitrogen since 70 per cent of grain-N is derived from disassembly of leaf- photosynthetic protein during senescence, results in increased grain growth hasten canopy senescence and finally restricts grain feeling duration.

Rice grown under elevated CO_2 concentration increased the ethylene production which may enhance tillers production and release of auxiliary buds leading to increase in grain yield. However, in some other studies under laboratory conditions though, doubling in CO_2 concentration accelerated the vegetative growth but increase in grain production was not recorded due to simultaneous increase in temperature leading to increase in photorespiration. Therefore, it is very difficult to derive any final conclusion on crop benefited with increase in CO_2 concentration without combined effects of enrichment in temperature, nutrients and moisture supply.

15.1.5 Effect on Yield and Quality

In discussion on the effects of elevated $[CO_2]$ on photosynthesis, the rate of this process increases more rapidly in C_3 species. It also helps in opening of stomata and thus increases water-use-efficiency in both C_3 and C_4 plants results increases in yields. Influence of elevation of $[CO_2]$ on photosynthesis and yields were more pronounced in C_3 species than C_4.

Response to elevated $[CO_2]$ on grain yields also varies among varieties of the same crop which has been widely noticed in rice followed by soybean cultivars and some other crops.

In several experiments, even 1,00 to 1, 500 ppm of $[CO_2]$ increased the yields of C_3 grain crops but unfortunately, the simultaneous effects of rise in temperature and $[CO_2]$ under field conditions has not been observed. Benefit in yields obtained from elevated $[CO_2]$ may be nullified by increase in the canopy temperature and gain in net-energy harvest may be ultimately same.

In another study, doubling CO_2 concentration projected 50 per cent increase in yield of soybean but a 3°C rise in surface air temperature almost set off the positive effects of doubling CO_2 concentration.

Enrichment of $[CO_2]$ affects amino acids transport from leaves to grains and hence the grain nitrogen (protein) concentration decreases. The changes in grain –N concentration caused by increased $[CO_2]$ may vary from one species to another. As such it is not affected in rice and soybean but grain quality in wheat, barley and cotton reduced significantly. Elevated CO_2 concentration also altered the wheat grain lipids and doubled the number of mitochondria in leaves, compared to the ambient $[CO_2]$.

15.2. Crop Response to Elevated Temperature

15.2.1 Effect on Vegetative/Reproductive Growth

Warming in tropics may compel species/genotypes selection difficult. Even shifting in crop and cropping systems are possible. Rise in temperature in the temperate world may shift the cultivation towards the pole for some species. Present rise in temperature has already 10-15 per cent globally reduction in wheat productivity and similar to maize and rice have already been recorded.

High temperature may fasten vegetative growth but reduces vegetative phase and brings earlier reproductive phase. It has marked injury to ovule and flower numbers, pollen survival, pollination, fertilization and finally grain size or test weight,

Among all, pollen is the most susceptible to rise in temperature which is adversely affected even rise in critical temperature (30^0C for wheat and 34^0C for rice). In maize, temperature above $32°C$ brings out complete sterility and loss in grain setting. At high temperature several pollen enzymes which are essential for carbohydrate metabolism becomes inactive. Such damage occurs during meiosis of microspore cells or after meiosis around the time when pollen microspores are released to form tetrads.

High temperature interferes with hormonal signaling from the embryosac, which controls the direction of pollen tube growth. High temperature at early seed development stage can result to ovules abortion as such first three days elevated temperature ($>30°C$) affects up to 70 per cent reduction in grain number in wheat.

Stigma and style are less sensitive to temperature compared to anthers, pollens and ovules. In rice, the upper critical temperature for normal functioning of stigma and style is at least $6°C$ higher than for pollen. Therefore, sensitiveness to higher temperature of pollen is more accountable for reduction in number of grains.

15.2.2 Effect on Grain Feeling Duration, Grain Feeling Rate and Yield

In wheat, grain feeling duration is reduced by 3.1 days for each degree rise in temperature. Decreases in grain feeling rates at high temperature, are due to specific effects on the enzymes taking part in conversion of assimilates provided by the mother plant into seeds. Temperature above 30°C, enzyme like soluble starch synthase is a key enzyme in amylopectin production, is very sensitive. Amylopectin which is very sensitive to higher temperature, contributes for 50-60 per cent of the grain weight of three major cereals; rice, wheat and maize.

In addition to soluble starch synthase, elevated temperature during kernel development after 10-15 days of pollination can inhibit cell division of endosperm tissue and restrict the dry matter accumulation in grains.

Relatively high temperature generally has adverse effects on grain yield. As such, the grain yield of wheat can decline by 5 per cent for each 1°C rise in temperature from 17.7°C to 32.5°C and rice grain yield by 4.4 per cent with rise in temperature from 26.7°C to 35.7°C. In maize, extreme summer temperature can result to complete restriction in grain formation. Even high night temperature is injurious for grain feeling in sorghum. Elevated temperature is more deleterious for pulses than cereals. Grain yield in bean can be 11 per cent less to each degree rise in day time/night time temperature combination of 28/18°C under controlled conditions while increase in temperature above a mean growing-season temperature of about 20°C under field conditions.

According to some works reported in India on the effect of rise in air temperature, 2°C increase in mean air temperature would decrease rice yield by 0.75 t/ha in high yielding areas and 0.06 t/ha in low yielding coastal region.

Only 0.5°C increase in winter temperature would reduce wheat crop duration by 7 days which will reduce yield by 0.45 t/ha while 0.5°C increase in water temperature will reduce wheat yield by 10 per cent in wheat bowel area and by 7.9 per cent in other parts.

From this projection (Table 15.1), it appeared gradual decreases in wheat yields at each degree increase in temperature at present CO_2 concentration (350-387 ppm) is possible but increase in yield at 550 ppm of CO_2 up to 3°C increase in temperature may be contradictory since photosynthetic apparatus in most of the plants in general and C_3 in particular ceased to function above 450 ppm CO_2.

Table 15.1: Projected Decreases/increases in Wheat Yield at Each Degree (°C) Increase in Temperature at Two CO_2 Levels.

Temperature (°C)	Yield decrease (per cent) at 350 ppm CO_2	Yield Increase/Decrease (per cent) at 550 ppm CO_2
1	− 5	12
2	− 12	7
3	− 21	1
4	− 25	− 5
5	− 31	− 11

In another report reduction in yield of wheat, rice, maize and groundnut was recorded due to increases in temperature from 1 to 3°C (Table 15.2).

Table 15.2: Reduction in Yield at Increased Temperature.

Increase in Temperature (°C)	Wheat	Rice	Maize	Groundnut
1	8.1	5.4	10.4	8.6
2	18.7	7.4	14.6	23.2
3	25.7	25.1	21.4	36.2

According to another projection, the impact of rise in 2°C temperature and 7 per cent in precipitation may not be positive for crops due to loss in total farm produce caused by erratic rainfall pattern, flood and drought. The effect of temperature will be more adverse for wheat production in wheat dominant states and coastal Tamil Nadu but it may be beneficial for Eastern UP, Bihar and West Bengal.

15.2.3 Effect of Elevated Temperature on Grain Quality

Since high temperature can adversely affect the processes occurring in grain feeling hence it will result in small shriveled grain formation and poor milling and cooking quality. Grain protein in most of the crops *viz.*, soybean, maize, rice, wheat and barley increase with rise in temperature. However, in some others sensitive wheat varieties, protein content deteriorate at temperature above 25°C and in resistant one above 35°C. In barley too, malt quality decreases in grain formed at high temperature.

Disruptions in growing grain carbohydrate metabolism are recorded at high temperature in some major crops; rice, wheat and maize. Wheat and rice grain formation at high temperature has led to chalky appearance attributed to small starch granules surrounded by several air spaces. In rice starch amylase concentration decreases to 0.4 per cent for every 1°C rise in temperature during 5-15 days after anthesis, causes in poor cooking quality like loss in expansion and increases in stickness. Similar results are also noticed in baking quality in maize.

In oilseeds crops, in sunflower, high temperature significantly decreases the oil quality due to shortening in ratio of linoleic acid to oleic acids from 6:1 at 12°C and 1:1 at 28°C.

15.2.4 Plant Adjustment to Temperature

Some of the plants have their photosynthetic adjustment characteristics to sustain under high temperature due to

1. Plant inherent genetic diversity
2. Plant with differential strategies in growth and development
3. Plants respond to temperature changes rather than to absolute temperature.

Therefore, high and low temperatures are relative terms which differ for pychrophilic, microphilic and thermophilic organisms. Accordingly, at low temperature (0–10°C), the photosynthesis is constrained thermodynamically while at high temperature (35 – 50°C), the thylakoid membrane stability limits the photosynthetic performances.

At high temperature quenching of excitation energy decreases while photorespiration increases, which results in decrease in the quantum yield of CO_2 assimilation due to down regulation of Rubisco by 50 per cent at temperatures 30 – 45°C.

Though, even tropical plants may survive below 40-44°C temperature but some of the desert species like *Tidestroma oblongifolia* may continue photosynthesis when exposed to 44-50°C while *Stripa* spp., *Corex humulus* and *Bothriocloa ischaemum* can survive at 65-70°C temperature.

Some bacteria (*Themotoga* spp.) can grow in thermal spring at temperature 90°C. Members of the genus *Pyrodictium* grow at 110°C in the field of hot, bubbling sulfurous mud (sulffatara fields) in the hydro thermal system. *Isola volcano* due to having some special lipids in its membranes and their DNA is protected by special histone like proteins.

15.3 Effect of Elevated Ozone (O_3) on Plants

Higher concentration of ozone in lower sphere oxidizes protein and lipids of leaf membranes, increases their permeability and lowers the photosynthetic capacity. This also adversely affects short and/or long distance transport of metabolites in whole plant body. [O_3] often causes immature leaf-senescence and thus reduces leaf area and leaf duration.

Exposure to high [O_3], limits the source strength by directly damaging the photosynthetic apparatus as well as weaken the long term whole plant source strength by reducing leaf area. It also inhibits the plant mechanism to detoxify and repair the damaged tissues. Therefore reduction in substrate reduces cell division and cell expansion.

15.3.1 Effect on Growth

Ozone pollution leads to decrease in total dry matter weight of plant shoots but high variations in leaf area expansion and entire plant leaf area were also recorded. Plants grown under elevated [O_3] conditions produced more leaves to compensate shorter leaf duration. Radish exposed to high [O_3] grew leaves faster than in its absence, at the cost of roots and stems.

At high [O_3], total root bio-mass always decreased. This may be due to the fact that under resource stress conditions roots allow the resources to move towards shoots to maintain above ground development.

15.3.2 Effect on Reproduction

Pollen germination and pollen tube growth are very sensitive to high [O_3]. This might be due to directly reactivations or indirectly in activation of a specific regulator of pollen tube growth in presence of high [O_3]. The influence of O_3 on pollen germination and tube growth may also be responsible for change in the stigma instead of the pollen itself. Since pollen germination occurs on the stigma, while pollen tube growth occurs within the style and therefore, these processes may have different responses to elevated [O_3] as stigma is protected from direct exposure to ozone. Therefore, elevated [O_3] is more injurious to flowering and physiological maturity.

Special sensitivity to elevated [O₃] around the time of anthesis itself indicates that sexual reproduction period is more vunerable than any other stages.

15.3.3 Effect on Grain Feeling Duration and Rate

In several crops reduction in grain or fruit size has been found with a maximum in soybean and wheat due to shortening in both duration and rate of grain feeling stages. Exposure of crops to elevated [O₃] commonly results increases in grain protein and concentration of minerals but in oil seeds crops a clear decrease in oil content was found.

15.3.4 Effect on Yield

Even present days tropospheric ozone concentration has limited the yield potential of several crops and therefore, further increases in [O₃] may damage a number of other crop species. A negative linear relationship between yield and [O₃] in many crops has been recorded. Since ambient O_3 concentration is more temporally and spatially variable than CO_2 concentration and therefore, it is difficult to predict a normal troposphere [O₃] effects on crops in general.

In cotton, depending on cultivars,(Figure 15.1) a field loss from 0-40 per cent at high [O₃] was recorded In C_3 grain crops also negative effects

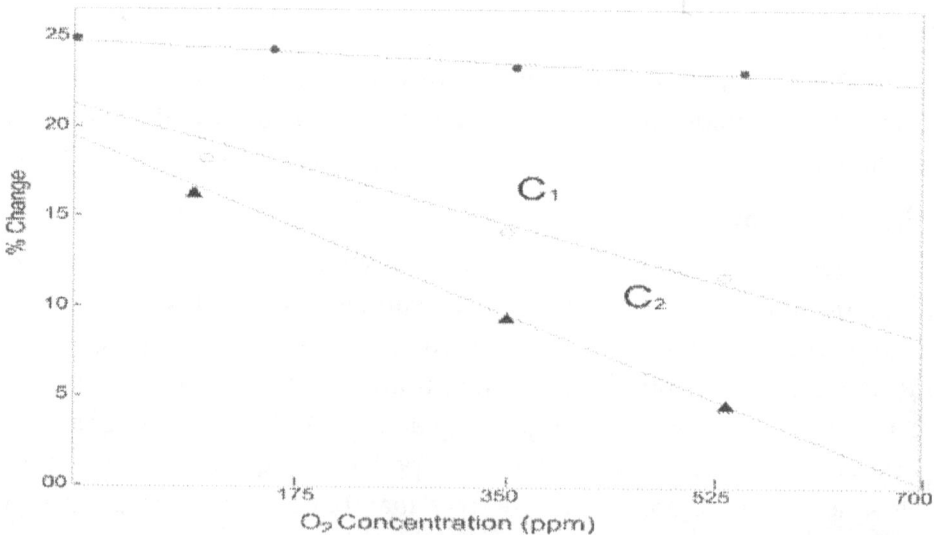

Figure 15.1: Decreases in Photosynthesis, Carbohydrate Content and Export to Elevated O_3 for 0.75 hr. in Two Cotton Cultivars (C_1 and C_2).

on yields as compared to C_4 (maize and sorghum) were observed. This was possible since the adverse effects on yields components were equally pronounced.

Specially, Crops exposed to $[O_3]$ before anthesis were more adversely affected due to reduction in grain number. This was due to inhibition in photosynthesis, premature leaf senescence, loss in plant canopy leaf area and distribution of assimilates to active sinks.

15.4 Interactive Effect on Yield

Simultaneous effects of changes in different climatic parameters are more accountable than effects of sole factor. Several reports suggest that increases in $[CO_2]$ will accelerate photosynthesis without considering the elevation of temperature which may ultimately increase respiration and thus loss in net-assimilation. Studies under laboratory controlled conditions on the effects of one parameter may not be equally good under open field conditions. Till date very few reports are available on the interactive effects of two or more than two indices under field conditions.

15.4.1 Effect of Elevated CO_2 and Temperature on Yield

The combine effects of elevated $[CO_2]$ and temperature compared to ambient $[CO_2]$ and temperature, reduced the yield in a dozen of experiments while about increases in yields in a half experiments. This has indicated that stimulation in yield may be nullified by simultaneously increases in temperature.

In wheat, an increase in the mean temperature of 0.7 to 2.0°C during the growing season was enough to reduce the increase in yield recorded due to doubling the $[CO_2]$. In one of the study conducted in India 17 per cent reduction in yield due to increase in temperature by 2°C and beyond this level, decrease may be even more. Elevation in $[CO_2]$ can be compensated to a certain level due to positive effects on crop growth. 450 ppm $[CO_2]$ can offset the negative effects of 1°C rise in temperature and for further 550 ppm $[CO_2]$ may be further needed to nullify the negative effects of 2°C elevation in temperature. However, simultaneous increases in UV-B radiation and $[O_3]$ may create problem in wheat production in present day wheat growing area.

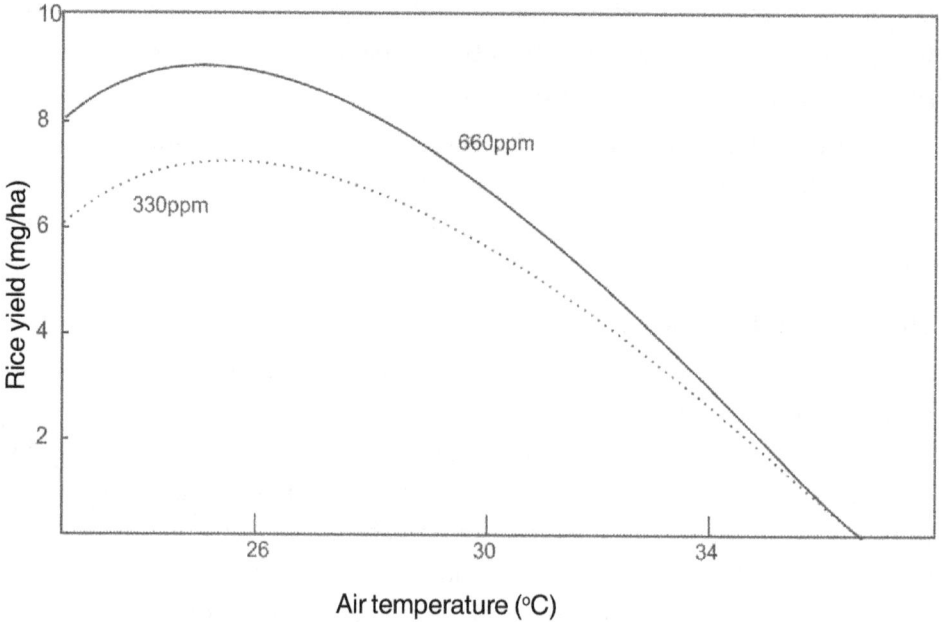

Figure 15.2: Effect of Temperature on Rice Yields Influenced at two [CO$_2$]. Similar Trends in Yield Reductions in Presence of both [CO$_2$] after 25°C was Quite Apparent.

In rice (Figure 15.2), at temperature about 27.5°C has eliminated the effects of doubling concentration of [CO$_2$] on dry matter and ear head production.

In soybean, doubling [CO$_2$] may increase the yield by 50 per cent but simultaneous increases in maximum and minimum temperature to 1°C and 1.5°C, the grain yield may come down to 35 per cent.

Since elevated CO$_2$ causes closure of stoma, canopy temperature may increase in CO$_2$-enriched climate due to reduction in evaporative cooling. This confirmed that the elevated CO$_2$ and increase in temperature are directly related to increase the degree of sterility in crops including rice.

15.4.2 Effect of Elevated CO$_2$ and O$_3$ on Yields

Crop yield losses due to elevated [O$_3$] may not be completely encountered by elevated [CO$_2$] since it cannot protect the crops from direct damage of [O$_3$] on reproductive phase starting from pollen germination to pollen tube elongation. However, elevated [CO$_2$] may reduce the injury of [O$_3$] by reducing the stomatal opening and thus restriction in entry of

O_3 into leaf cells to some extent. Thus O_3 polluted area is of greater concern to CO_2 polluted one at least from crop production point of view On the other elevated [O_3] results in higher protein concentration in green while elevated [CO_2] gives higher carbohydrate yields and therefore, combination of both may result a modest increase in grain protein.

Studies on the combined effects of elevated [CO_2] temperature and [O_3] particularly under field conditions are yet lacking. Therefore, It is very difficult to derive a concrete conclusion on increase or decrease in yields of different crop species which may even vary from one region to another. However pollution at least will change the crop and cropping system from one region to another. A shift in better crop performance towards the pole may take place or say temperate may change to sub-temperate.

15.5 Crop Response to Enhanced UV-B Radiation

Increase in UV-B radiation at the earth surface due to depletion of stratospheric ozone layer will affect DNA duplication and gene transcription which will bring out changes in structure and functions of cells. This will affect the production potential of several crops.

Rice

Rice (*Oryza sativa*) genotypes are very sensitive to UV-B radiation due to lacking capacity to repair the damaged DNA. Plants with steeply inclined leaves minimize the UV exposure by reflecting or absorbing UV in the epidermis. These cells due to containing phenolic compounds absorb UV-radiation and do not allow UV-radiation to penetrate the epidermis beyond 32mm which usually occurs up to 75 mm penetration in non-phenolic compounds producing species, like rice. However, in one of the laboratory experiment, the UV-B radiation showed reduction in chlorophyll content of the leaves at tillering stage but it recovered in new growth tissues in super- high yielding hybrid rice. The light saturated photosynthetic activity of UV-B exposed plants was however higher over control.

Wheat

Some of the works done on the performance of wheat under elevated UV-B radiation indicated up to 66 per cent decrease in net photosynthesis.

Summarizing the other works reported in respect to effects of UV-B radiation on different aspects of plants performance it appeared that

☆ Species originated from locations of high total solar radiation as of Mediterranean area will be rather benefited by increase in UV-B radiation.

☆ Elevated CO_2 induced increased assimilation and water use-efficiency but raised UV-B radiation lowered the transpiration rate

☆ Enhanced UV-B radiation adversely affects the yields of wheat, pea plants, bush bean, rice, soybean and cotton by 10-50 per cent but it may increase the bio-mass production of barley and strawberry.

15.5.1 UV-B Radiation and Nutrients Concentration in Plants

1. *Brassica napus* grown under UV-B radiation along with high concentration of cadmium, the Mg concentration in plant reduced but concentration of Mn, Ca, Cu and K were increased.

2. High level of N increases the effect of UV-B radiation.

3. According to one study, balanced fertilization of N, P, and K ameliorates the adverse effects of UV-B radiation in wheat and *mung*/green gram.

4. UV-B radiation has bad effects on a number of enzymes.

5. As reported, Selenium spray reduces the effects of UV-B radiations.

Table 15.3: Plants Sustain to Low and High Irradiance.

Species	mmol Cytochrome f Content per Unit Chlorophyll
Spinacea oleracea	1.20 – 5.00
Ipomea pentaphylla	0.74 - 4.05
Phaselous mungo	2.08 – 3.88
Sinapis alba	1.00 – 3.57
Tradescanta albiflora	0.42 – 1.25
Alocasia species	0.70 – 1.25

15.5.2 Radiation and Chlorophyll

Cytochrome f content to per unit area of chlorophyll in leaves of the species indicates their potential for growing in presence of high and low irradiance (Table 15.3). Plant species, *Spinacea oleracea* and *Phaseolus mungo* are better adapted to high irradiance while *Tralescanta albiflora* and *Alocasia* sps. are sensitive to high irradiance. Wider range in Cytochrome f content in *Ipomea pentaphylla* has made it to sustain under both low as well as high irradiance conditions.

15.6 Cold Stress and Plant

Low temperature is one of the most abiotic stresses that affect temperate plants since water and temperature are the major determinants of plant life. Polar region and Oceans with low temperature occupy about 80 per cent of the Earth's surface. Only one-third of the total land is free of ice and 42 per cent of this land has temperature below – 20°C. Majority of temperate crops are sown in the autumn and complete maturity in coming summer. Survival of the plant over winter period is termed as winter hardiness or frost hardiness. Identification of the key genes underlying cold stress will help in breeding of cold resistant cultivars. Cold-adapted plants tend to be slow growing, have the C_3 mode of photosynthesis and store sugars in underground tissues Plants adapted to cold climate have an efficient respiration mechanism, controlled by genes, which allows them to rapidly mobilize stored material during the short growing season.

15.6.1 Stress Avoidance and Tolerance

Stress avoidance and tolerance are the two mechanisms by which the plants sustain under low temperature. Stress avoidance protects the tissues from freezing. Some succulent species with thick tissue mass and sufficient water content are able to accumulate residual heat during the day and dissipate it slowly in cold at night while a number of herbs survive in form of dormant seed and protecting shoot meristem with leaves. Sometimes when temperature goes down to – 40°C, the endogenous ice nucleation is prevented by inhibiting the formation of ice nucleates. Some extremely winter hardy species survive even at – 196°C by generating a high viscous solution **"liquid glass"**, that prevents ice nucleation. These cells become osmotically, themally and mechanically de-sensitized to the presence of external ice.

The plant response to low-temperature stress has three distinct phases.

1. Pre-hardening that occurs at low, temperature but above 0°C

2. Hardening at which the full degree of tolerance is achieved, requires exposure to a period of sub-zero temperature and

3. Recovery of plant after winter.

15.6.2 Cold Acclimation

Overwintering temperate plant species acclimatize in winter during which their metabolism is redirected towards synthesis of cryoprotectant molecules, soluble sugars, sugar alcohols and low molecular weight nitrogenous compounds. These, in conjunction with dehydrin proteins (DHNs), cold regulated proteins (CORs) and heat shock proteins (HSPs), act to stabilize both membrane phospholipids and proteins, and cytoplasmic proteins, maintain hydrophobic interactions and ion homeostasis and scavenge reactive oxygen species (ROS). Some other solutes released from symplast serve to protect the plasma membrane from ice adhesion and subsequent cell disruption. Symplastic and apoplastic soluble sugar – not only fructan precursors, but also trehalose, raffinose and some others contribute to membrane stabilization.Increased activities of other antioxidants also accelerate the processes.

15.6.3 Cold Sensing and Signaling

Plant sensors of low temperature include Ca^{2+} influx channels, two-component histidine kinase and receptors associated with G-proteins. Certain cytoskeletal components participate in cold sensing by modulating the activity of Ca^{2+} channels following membrane rigidification. Therefore, ROS, Ca^{2+} - dependent protein kinase (CDPKs), mitogen-activated protein kinase (NAPK) cascades and the activation of transcription factors (TFs) promote the production of cold response proteins. These products can be divided into two distinct groups:

1. Regulatory proteins controlling the transduction of the cold stress signal

2. Proteins functionally involved in the tolerance response

15.6.4 Enzymatic and Metabolic Response

Several enzymes are involved in the cold response machinery. In addition to those associated with osmolyte metabolism, detoxification cascades and photosynthesis, lignin metabolism, starch metabolism and some others participate to cold stress, while the genes are involved in photosynthesis, cell wall, lipid and tetrapyrrole synthesis.

For example the drought, salinity and cold tolerance of rice transformed with an over expressed *Escherichia coli* trehalose biosynthetic gene were all significantly wild than the wild type. In wheat and tomato invertase activity is up-regulated by a fall in temperature. In addition to these enzymes, metabolism of nitrogenous compounds, also play their role in reducing low-temperature stress in plants.

15.6.5 Sugar as Signaling Molecules

Sugars are not only a source of energy but act as carbon precursors, substrates for polymers, storage and transport compounds and signaling molecules. The extracellular sugar concentration in cold-induced barley cell cultures regulates expression of the stress-responsive genes BLT4.9 and DHN1. Three different signaling pathways are kinown in plants:

(i) Hexokinase-dependent

(ii) Glycolysis-dependent and

(iii) Hexokinase-independent

Hexokinase functions as an intracellular glucose sensor, while some membrane have receptors act as extracellular sensors.

Fructose-based polymers also help in cold and drought tolerance in several plant species by binding to the phosphate and choline groups of membrane lipids. help in reducing water loss. Fructans accumulation in wheat, barley, oats, Poa and Lolium helps in reduction of water from the tissues but not in rice and maize.

Chapter 16
Biotic Stress and Crops

Effects on crops due to interactions with weeds and pests in changing climate are of vital importance for managing the crops and cropping systems in present day cropping system. Since crops and weeds compete for the same space, water, nutrients and sunlight for their growth and development hence competition is a must. Majority of the grasses belong to C_4 species while major grain crops are the C_3 species hence C_4 grasses mostly present in tropics are the dominant species over C_3 in terms of both underground and aboveground competitions for the same resources.

Well-developed adventitious roots of C_4 grasses are more efficient in extraction of soil moisture and nutrients due to their higher root-cation-exchange-capacity as well as their ability to sustain under stress environmental condition than crops. Some of them are also efficient to grow in problematic soils conditions and to change the unavailable form of nutrients into available form. Low threshold values for several essential nutrients further make them as a dominant species in association with crops.

Fast germination and emergence capacity facilitates the C_4 grasses to compete for space and light to dominate over crops. Their survival under high light intensity of shorter wavelength, high temperature conditions as well as better efficiency to tolerate the drought spell compared to C_3 crops make them in advantageous position.

16.1. Effect of Elevated Temperature on Weeds

The effect of elevated atmospheric temperature on crop x weed association appeared to be dependent on the geographical situation, origin of the species and the crop grown.

Since majority of the weeds are the habitat of tropical and warm-temperate climates irrespective of their photosynthetic pathways either C_3 or C_4 species hence, infestation of these weeds towards the poles is expected as the temperature is rising.

Some of the studies on crop x weeds competitions under elevated temperature indicate that weeds are at the advantageous position. Elevated temperatures (26/17, 29/20 and 32/23°C day/night temperatures) help weeds to suppress soybean growth. Increase in temperature from 26/22 to 34/30°C day/night also resulted weeds as a aggressive species over crops.

In some of the other studies, the effects were in reversed order as warming reduced the weeds population. In cotton, a C_3 tropical crop gained competitive advantage over weeds at 32/23°C compared to 26/10°C day/night temperature.

16.1.1 Effect of Elevated [CO_2] on Weeds

If elevation in temperature favours C_4 species, the enhancement in [CO_2] helps C_3 species. Therefore, increases in atmospheric CO_2 logically may be favorable for C_3 crops over C_4 weeds. However, if the crop and weeds both belong to C_3 category, the weeds may be at advantageous position due to their other competitive capability and hence, warming in any case may be injurious to crops.

In one of the experiment wheat grown at double to ambient [CO_2], the grain yield was increased by 30 per cent but when grown with weeds, the yield was reduced. In soybean, 25 per cent increase in yield was recorded at elevated CO_2 in weed free plots, compared to weed infested plots.

At 720 ppm CO_2, the growth of weed was stimulated to that grown at 360 ppm CO_2. The weed was even resistant to a popular weedicide, Roundup (Glyphosate). This might be due restriction in entry of weedicide to reduce stomatal aperture in presence of elevated CO_2. Reduction in transpiration rate under elevated CO_2 may also reduce the effectiveness of the weedicides. Even higher $[CO_2]$ stimulates the underground weeds tubers and rhizomes to increase in their reserve materials to enhance weed regeneration capacity.

16.1.2 Effect of Elevated $[O_3]$ on Weeds

A very few studies on the effects of elevated $[O_3]$ on crop x weed association have been reported which are even limited to pasture species. In grass-legume association, grasses are benefited under higher ozone concentration. Besides several other reasons, this might be one of them for aggressiveness of grass component in pasture system. Sometimes, it has been also observed that at elevated $[O_3]$, one C_3 weed may disappear while others C_3 may come up.

16.1.3 Effect of $[O_3]$ on Insects and Diseases

Changes in climatic conditions may affect on the incidences of insects and diseases on crops by;

1. Modifying the chemistry of plant trissue by changing their nutritional status and properties.
2. Speeding or slowing the development of crops thereby synchronizing the timing of crop susceptibility to pests or
3. Altering the developmental rate or demography of pests themselves, and thus changing the timing of pest attacks.

Global heating may bring climatological changes to influence the crop-pest interaction due to changes in their physiology and nutritional quality. The effects of change in climate on biological defences of crops may be possible. The alkaloids, terpenoids and phenolics are the three main groups of protective chemicals which act as a defence. Alkaloids are the nitrogenous compounds, synthesizes from amino acids while terpenoids are the carbon based molecules (lipids) syntheses via the mevalonate pathway and phenolics are the aromatic carbon-based molecules formed via the shikimate pathway. Among these three, the alkaloids are less sensitive to climatic fluctuations compared to other two.

Some of the crops produce different chemicals to protect themselves from pests and diseases. DIMBOA (2,4-dehydroxy-7-methoxy-1,4-benzoxazin-3-1) produced by maize plant is a natural pesticide with effectiveness against some borers, bacteria and fungi. Cabbage and cassava also synthesize glucosinilates and cyanogenic glycosides, respectively as protective chemicals against pests and diseases. Sorghum, potato and cotton also produce some amount of phenolic compounds which have antiherbivore properties. Some crops of *Solanaceae* family; tomato, potato and pepper also contain alkaloids which are toxic to insects. Hence, it is high time to record the changes in climatic conditions on the production of these three defensive chemical compounds and to transfer the same gene to other crop species as a tool for self-defense.

16.1.4 Effect of Elevated Temperature on Pests

Lower temperature in winter season results in less incidences of pests and diseases while high temperature and high humidity in rainy season are favourable for incidences of pests and diseases. It seems that elevation in temperature to just a few degrees may increase the population of pathogens very quickly. This may be more apparent during reproductive phase of the crops. It is also true that higher temperature the insects in early completion of their life cycle. Only 2°C rise in temperature helps several insects to complete their life cycle by 2-3 weeks earlier. Warming also advances flight phenology of aphid species by more than a month.

Global warming enables a number of pests in intensification of their geographical distributions. This new happening will experience the incidences of new diseases and pests in un-conventional areas. Fortunately, elevation in atmospheric temperature will be also conducive in increasing the population of beneficial insects which will help the crops in pollination. Simultaneous increases in those birds which feed on insects may also increase to benefit the crops.

16.1.5 Effect of [CO_2] on Pests

Higher [CO_2] may decrease the tissue amino acid content as well as total-N concentration associated with higher C:N ratios. This will help the insets in easy feeding of succulent tissues as much as to 40 per cent. On the other hand insects reared on low protein leaves may suffer since crop grown at higher [CO_2] are less nutritious.

Increases in phenolic compounds in plants further reduces the incidences of insects attack but chewing insects which are phloem feeders may be benefited by elevation in CO_2 concentrations.

Very scanty information is available on the effects of elevated atmospheric $[CO_2]$ on the incidences of plant diseases. Enrichment in plant-N invites insects for the damage but plant grown at high $[CO_2]$ concentration are low in N, does not attract the insects.

Elevation in $[CO_2]$ might be beneficial for soils pathogens by stimulation of root growth and root-exudation of sugar into soils. Such crop may be less susceptible to diseases. In one of the experiment, population of bacteria and actinomycetes in cotton leaves were unaffected at 550 ppm of CO_2 under field conditions. At the same time the population of some fungi were lower, some others were unchanged while some others were increased.

16.1.6 Effect of $[O_3]$ on Pests and Diseases

Crop grown in presence of elevated $[O_3]$, produces more succulent leaves due to wider C: N ratio, are congenial for chewing insects. Such plants favour the growth of insects and reduce their mortality, is very common in tobacco plants.

High $[O_3]$ can enhance the production of secondary compounds in some plants like signal molecules, PR protein and antioxidative systems. High $[O_3]$ also helps in the production of elicits cellular barriers like callose which lignify the cell walls to resist the attack of pathogens. Exposure to high O_3 further increases host plant susceptibility to necrotrophic and root-rot fungi. Elevated $[O_3]$ reduces the infection in soybean from *Pseudomonas glycinea* and to alfalfa from *Xanthomonas alfalfae* bacteria. Wheat grown at high $[O_3]$, resists the infection of *Puccinia graminis* to sub-stomatal mesophyll cells.

16.2 Climate Change and Soil Organisms

The changes in atmospheric $[CO_2]$, $[O_3]$, temperature, UV-B radiation and several others pollutants have the primary direct effects on crops and the effects of symbionts may be secondary.

16.2.1 Effect of $[CO_2]$ on Soil Organisms

Direct effects of elevated $[CO_2]$ on fungal growth hardly occurred since roots mycorrhizal colonization is normally unchanged in plant roots under elevated $[CO_2]$ however, fungal biomass yield increased with increase in root weight. It further depends on soils physical and chemical properties including soil moisture status. As such, in sorghum, total (*Arbuscular mycorrhizal*) fungal hyphal length in roots was increased to 109 per cent at 550 ppm [CO_2] compared to 370 ppm under proper moisture conditions, the same was 267 per cent under water stress conditions.

N-fixation is usually activated under elevated [CO_2] in many legumes; clovers, soybean, pea and others due to increases in population as well as size. Fixation rate may be high in absence of amide and ammonical forms of N since these forms compete with bacteria for soil oxygen. Population of free living rhizobia in soils also increases by [CO_2] in clover and increase in number of N_2-fixing bacteria in rice soil was also stimulated due to flow of higher carbohydrate from shoots to roots.

16.2.3 Effect of Elevated Temperature on Soil Organisms

Effects of elevated temperature on *Arbuscular mycorrhizal* (AM) fungi is not recorded. Though, increases in fugal colonization in barley roots which was recorded at 10°C temperature, reduced at 15°C however it was increased in *Plantago lanceolata* weed at 20°C compared to 12°C. In some forage crop too, increases in fungal colonizations were recorded by 15 per cent when temperature was raised by 4°C.

Higher root temperature may reduce the N-fixation. In temperate legumes (clover, pea and bean) the N-fixation was reduced at 30°C and in tropical legumes (soybean, guar, peanut, greengram and cowpea) above 35-40°C. Low temperature may also reduce N-fixation therefore, warming is detrimental for N-fixing bacteria in tropics but it may be beneficial in temperate regions.

16.2.4 Effect of Elevated Ozone concentration [O_3] on Soil Organisms

Increases in [O_3] in atmosphere may have adverse effects on soil organisms as it has been observed in roots of soybean and tomato. Reductions have been recorded in N-fixations in soybean and beans at high [O_3]. Though, the adverse effects of elevated [O_3] on N-fixing bacteria is not possible since ozone failed to penetrate into the soil however, its restriction to photosynthesis and transport of assimilates to roots, reduces the supply of energy to rhizobia.

Therefore, rising [CO_2] accelerates the growth and activity of both *mycorrhizae* and *rhizobia* but this can be counteracted by [O_3]. Exactly, rising temperature may increases insects population which can be restricted by [CO_2].

Finally, one cannot reach to a final conclusion on these aspects due to availability of scanty information on the interactions effects of temperature, CO_2, UV-B radiation, O_3 and other environmental indices.

Chapter 17
Mitigation of Polluting Agents

In general, deforestation and changes in land use system is the source of increase in CO_2 concentration while rice – wheat system has contributed in production of CH_4 and N_2O in particular and animal production is the source of CH_4 emission. Global emission of non-CO_2 greenhouse gases is more than 5969 Mt equivalent to CO_2/year of which Agriculture shares 14 per cent in CO_2, 47 per cent of NH_4 and 84 per cent of N_2O productions (US, EPA, 2007).

The total anthropogenic greenhouse gases emission in 2004 in terms of CO_2 equivalent was the highest through energy supply (Table 17.1).

Table 17.1: Sources of Green House Gases Emission.

Source	Contribution (per cent)
Energy supply	25.9
Forestry	17.4
Agriculture	13.5
Transport	13.1
Building	7.9
Waste and Waste water	2.8

Mitigation of greenhouse gases in agriculture can be grouped in to three broad categories on the following principles.

1. Reducing emission
2. Enhancing removal of carbon through photosynthesis and returning the same in soils
3. Avoiding emission

17.1 Enhancing the Concentration of Soil Organic Matter

1. Minimum tillage to avoid burning of organic matter through sun light and return of residues to soils.
2. Improvement in water management to control burning of soil organic matter.
3. Growing and incorporation of thick canopy crops.
4. Use of bio-energy/bio-fuels due to C-neutral helps in reduction of CO/CO_2 gases in the atmosphere.

17.2 Methane Control

1. Production of rice under drying and wetting helps in reduction in emission of CH_4 gas.
2. Application of decomposed manures minimizes the production of this gas.
3. Uses of methanotropic bacteria in soils consume the CH_4 gas.
4. Mixed application of NH_4-fertilizer with organic matter reduces production of CH_4.
5. Addition of oils in feeds and feeding of improved forages also helps in reduction in formation of this gas.

17.3 Nitrous Oxide Control

Application of N- fertilizers and un-decomposed animal manures are the main sources of N_2O emission in to the atmosphere.

1. Covered manure heap can reduce N_2O emission to a greater extent.
2. Application of amide and ammonical forms of fertilizers under anaerobic and nitrate forms under aerobic conditions may help in mitigating N_2O emission.

17.4 Saving the Ozone Layer

Some of the possible measures if, adopted in daily life can help in saving the ozone layers are;

1. Aerosols containing CFC, fluorocarbon of CFC propellants should not be used.

2. Use of ozone – friendly aerosols such as those with hydrocarbon propellants or prefer to the brands that use pump spray containers.

3. Rejection of freezers, air-conditioners using Freon-11, Freon-12, Freon-22 and alike propellants can be a very positive measure in reduction of these longer life span injurious chemicals.

4. Non-use of halogen- filled fire extinguishers will also help reduce toxic chemicals.

17.5 Crop Management in Changing Climatic Conditions

17.5.1 Climate Change and Impacts on Crops

Drastic changes in climatic conditions leading to unpredicted drought in previously heavy rainfall areas on one side and extreme rain causing unprecedented flood in previously drought prone areas have virtually altered the ecology of the world. As a result;

1. Frequent occurrences of irregular monsoon and untimely rainfall may change the cropping pattern

2. Increased river flow and inundation during monsoon

3. heavy rainfall in short period results in water logging

4. Weak, late and short arrival in monsoon

5. Crop damage due to flash flood

6. Frequent drought in some parts while drought in some others

7. Reduction in rainy days and total rain in arid and semi-arid areas

8. Crop failure due to prolong drought

9. Some time short winter and some time prolong cold spell

10. Summer becoming more hotter

11. Salinity intrusion along the coastal region.

17.5.2 Plants Characteristics to Sustain Under UV-B Radiation

The plants characteristic which stand true to sustain under arid or desert conditions are suitable for survival under UV-B radiations.

1. Plants with erect leaf structure (erectophyll) are more effective than broad horizontal structure (planofix).
2. Secretion of gummy materials on leaf blade restricts transpiration.
3. Hairy leaves due to extension in their boundary layer also reduce transpiration.
4. Leaf with thorns and less in number with shrunken stomata also help in restoring water in plants.
5. Leaf thickness with stomata guided with double layers of guard cells may be a protective mechanism to sustain under shorter wave length of radiations.

Some of the works done on mitigating of the adverse effects of UV-B effects suggested that;

a. Plants treated with selenium increases the tolerance to enhance UV-B radiation due to its anti-oxidative action.
b. Production of **phenolic** substances and wax by some plants, help in screening and reflection of UV-B radiation and thus reduces damaging effects of high intensity radiations.
c. Production of antioxidative enzymes protects plants against the oxidative damages caused by UV-B radiation.

17.5.3 Global Warming and World Food Security

The UN Frame work Convention on Climate Change (UNFCCC) made two stipulations relevant to agriculture,

a. Prevent dangerous anthropogenic interference with the climatic system
b. To ensure that the food production is not threatened.

Global warming could affect food production in a number of ways to threaten food security for the world's most vulnerable continents in general and countries in particular. As such, wheat productivity in India may decrease to 60 percent by 2070 to 2,000 level. Shifting in monsoon may affect rice production in eastern and coastal India while pearl millet production in Rajsthan may reduce due to increases in temperature. An increase in CO_2 concentration may adversely affect soybean production in MP.

In Africa, increased drought could seriously affect millet yield by 63-79 per cent. In Middle East, rise in temperature could decrease wheat yield while decreases in yields of several major crops in Central America, Brazil, Chile and Mexico are expected. It will further decrease livestock population.

In China and Cooler parts of northern Asia, a marked decline in yields of rice, wheat and maize could take place but crops yields in northern Siberia may increase. Therefore, the following situations may arise.

- a. The success rate of predictability of weather and climate would decrease, thus making planning of farm operations more difficult which requires strong weather fore-casting system
- b. The rise in sea level, could threaten in submerging of coastal crop lands
- c. Biological diversity would be reduced mangroves and tropical forests
- d. Changes in world ecology may affect agro-ecological zones and press the farmers to change in their crops and cropping systems
- e. The marked changes in food production in tropical and temperate world could change the food habit
- f. Distribution and quantities of fish and sea foods may bring about changes in fishery industry
- g. Incidences in spread of diseases and insects in new areas may affect crop production

17.5.4 Global Warming and Possible Measures

The decline and degradation of natural resources ; land, soil, forests biodiversity and ground water due to excess exploitation will further deteriorate within the end of present century. Degradation of lands and fresh water scarcity in Africa, South Asia and some parts of Latin America has already posed a serious threat to mankind.

Some of the sustainable strategies that can reduce the emission of greenhouse gases to mitigate the climate change may be;

1. Adoption of cost effective energy-efficient technology in electricity generation, transmission, distribution and end-use can lower the costs and bring down pollution level besides reduction in emission of greenhouse gases since power generation sector is the major producer of these gases.

2. Shifting Cost-effective renewals enhances sustainable energy supply and reduces local pollution.

3. Forest conservation, reforestation, afforestation and viable forest management practices increase the sink capacity as well as protect watershed for increasing the ground water table and income.

4. Fast and reliable transport systems without or with minimum emission of injurious gases.

5. Uses of renewable energy like solar and wind energies in industries and transportation and households can reduce the emission of green house gases.

17.5.5 Global Warming and Future Research Thrusts

Some of the plant's characteristics may be helpful to identify and evolve new genotypes to enhance crop production in changing climate conditions.

1. Plants with carotenoids of the xanthophylls cycle are resistant to high temperature

2. Plants that can avoid the effects of UV-radiation

3. The xanthophylls cycle can prevent some of the potential damage by funneling off excess energy acting as a lightning rod, at both low and high temperatures

4. Isoprene production may provide additional protection of leaves at high temperature

5. Specific proteins and carbohydrates may also offer protection against higher temperature

6. Eco-physiological research and identification of high temperature tolerance gene may help in evolution of new eco-friendly genotypes.

7. Identification and isolation of the gene in leaves of *Quercus alba* which is responsible for sudden drop in leaf temperature to 8°C.

8. **Transgenic rice** plants bearing, the **cyclobutyl pyramiding** dimmers (CPD) **photolyase** gene of the UV-resistant rice cultivar has 5.1 to 45.7 fold higher CPD photolyase activities than the wild type, are the more resistant to UV-B induced growth damage, and maintains lower CPD levels in their leaves under elevated UV-B radiation.

9. In addition to identification and transferring of resistant gene from one crop to another, identification of such gene in other species needs attention. Conifer needles screen UV-B radiation far more effectively due to presence of absorbing compounds in the cell walls as well as inside their epidermal cells. The epidermis of the herbaceous species is relatively in-effective at UV-B screening because UV-B may still penetrate through the epidermal cell walls, even their vacuoles contain large amount of UV- absorbing phenolics.

Since, polyamines, waxes and alkaloids ether absorb UV-B radiation or act as a scavenger of the reactive oxygen species (ROS) hence, they are more resistant to UV-radiation.

References

1. C_3, C_4 Mechanism in Photosynthesis – Walky and Black

2. Climate Change and Crops – S N Singh

3. Crops and Environmental Change – Seth G. Pritchard and Jeffery S. Amthor

4. Ecology and Environmental Biology – Purohit/Agrawal

5. Environmental Science – (18 ed) Daniel D. Chiras

6. Global Climate Change – H D Kumar

7. Global Warming in 21st Century – K K Singh

8. Introduction to Plant Physiology (4rt ed.) – Willium G., Hopkins and Huner

9. Photosynthesis and the Environment (ed.) – Neil R Baker

10. Plant Physiology – Frank B Salibury and Cleon W. Ross

11. Pant Physiology (3rd ed.) – Taiz and Zaiger

12. Plant Physiological Ecology (2nd ed.) – Hans Lambers, Stuart Chapin III and Thijas L. Pons

13. Trace Elements in Soils and Plants (3rd ed.) 2001 – Kabata – Pendias

Index

A

ABA 63

Abiotic 2

Acids rain 104

Adaptation 14

Agriculture sector 103

Alkalinity 18

Aluminum 7

Antioxidation 63

Arsenic 109

Ash pollution 101

B

Biotic 2

Boron 22

C

C_3/C_4 122, 124, 125, 133

Cadmium 105

Calcicoles 23

Calcifuses 23

Calcium 26

CAM 127

Caparian band 29

Carbon sequestration 69

Climate change 147, 175

Climate stress 3

Cold acclimation 164

Cold stress 163

Crop 153

D

Dehydration avoidance 60

Dehydration tolerance 61

Desert 46

Development 150

D-genome 30

Diseases 161

Drought 46

Drought stress 44, 65

E

Effluents 102

El Nino 78

Euxerophytes 48

F

Fertilizers 101

Field capacity 44

Flooded conditions 39

Flooded soils 35

Food chain 104

Food chain 104

Food security 176

G

Genetic improvement 14

Germination 150

Global temperature 94

Global warming 62, 176, 178

Grain 153, 155

Grasses 33

Green house gasses 3, 92

Growth 150

H

Heavy metals 104

Herbicides 100

I

Insect 168

Iron 19, 24

L

La Nino 78

Lead 108

Leaf movement 53

Light 53, 138

Light utilization 52

M

Manganese 7, 22, 23

Mechanism of adaptation 37

Mercury 107

Mesosphere 117

Methane 174

Micronutrients 19

Mineral nutrition 39

N

Nickel 109

Nitrous oxide 174

Nutrient 17

Nutrient uptake 54

Nutritional stress 19

O

Organic acids 13

Ornamentals 34

Osmotic adjustment 61

Oxidation-reduction 39

Oxygen 144

Ozone 74, 83, 108, 141, 157 175

P

PAR 112

Permanent wilting 45

Pest 169

Pesticides 99

Phosphorus 11

Phosphorus transporter 13

Photosynthesis 50, 65, 122

Plant canopy 60

Plant performance 36

Plant phenology 64

Plant stress 11

Poikilohydric 48

Pollutants 3

Problematic soils 3

Q

QA 52

Quality 152

R

Radiation 3, 83

Radioactive pollutants 102

Rice 161

Root architecture 13

Root system 60

Ruderal species 5

S

SAR 18

Saturated soil 44

Sewage 101

SIC, 68

Soil acidity, 7

Soil conservation, 58

Soil organisms, 170

Soil pH 26

Spenders 47

T

Temperature 3, 52, 72, 82, 130

Thermosphere, 117

Tolerance, 14

Toxic wastes, 103

Transgenic rice, 179

Transperation, 52

Transport of nutrients, 39

Troposphere, 116, 146

U

UV- B 75, 112, 139, 163, 176

UV-A 112

UV-C 112

V

Vegetable crops 33

W

Water 31

Water conservation, 58

Water harvesting

Water logging, 40

Water pollution, 103

Water requirement, 81, 82, 83

Water stress, 46, 54

Weeds, 167

Wheat, 162

Wilting point, 40

WUE, 55, 64

Y

Yield, 151, 158

Z

Zinc, 22, 23

www.ingramcontent.com/pod-product-compliance
Lightning Source LLC
Chambersburg PA
CBHW021434180326
41458CB00001B/273